JOURNEYMAN ELECTRICIAN EXAM PREP

From STRESS to SUCCESS

Master Every Question with Comprehensive Walkthroughs and a Failproof Decoding Technique for Guaranteed First-Try Success

Max D. C. Crent

Download the BONUS content with the QR code at the end of the book

Table of Contents

PART I: SETTING THE STAGE FOR YOUR EXAM PREPARATION

READ THIS FIRST: How to Effectively Use This Book

Your money matters! Think twice before making significant expenses.

Get it into your head: to pass the exam, you don't need to buy super expensive courses and spend a ton of money! The key is learning how to find the right answers on exam day.

This book is an excellent starting point to build the confidence needed to read each question and identify the correct answer efficiently.

Definitions after definitions: your time is precious

The numerous Definitions within each chapter are aimed at simplifying the complex content of the NEC book to avoid you wading through over 900 pages of dense material. They will help you grasp the essentials quickly and efficiently and you will save a lot of time.

This book is not an electrical engineering course

This book is not intended to be a treatise on electrical engineering. The Journeyman Electrician candidate already has a basic understanding of electricity due to the significant number of apprenticeship hours required for taking the exam and obtaining the license.

This book is your way to pass the Exam

"Journeyman Electrician Exam Prep" book aim to help you build your solid method. It ensures that you, the aspiring Journeyman Electrician, could approach EVERY question with the certainty and confidence of finding the right answer. After each exercise, always ask yourself: "How did I solve it?". More than answers, try to find your METHOD!

Your GOAL is to pass the Exam and nothing else!

I can never stress the following concept enough: the core of your preparation must be centered around extensive practice! That will allow you to pass the exam.

At the end of each chapter, specific exercises are designed to test your understanding of the content and to build confidence in applying these principles. Completing these exercises is crucial for internalizing the chapter's material.

Furthermore, download the BONUS content available with the paperback version of this book. This additional material is designed to further enhance your skills, solidify your understanding of the methodologies introduced, and provide consistent approaches to tackling exam questions.

Introduction

Welcome to the journey of becoming a Journeyman Electrician! This book is here to guide you through the twists and turns of the electrical world, helping you understand the ins and outs of the trade. Before we begin, it is crucial to note that you need to have the NEC manual that is required for taking the exam in your specific State. Verify what is the current official edition. It is an essential resource for your work. All NEC code books can be freely read in the Free Access section of the NFPA website [https://www.nfpa.org/codes-and-standards/7/0/70]. After creating a free account, you can access the desired NEC Code book version. No download there. An alternative is here: [https://up.codes/viewer/minnesota/nfpa-70-2023]. It is recommended anyway to purchase the paper version as it will be required on the day of the exam. Make sure to have it handy throughout this adventure, as it will be a valuable tool in your learning process. Let's start then, and by the end, you'll be equipped with the knowledge and skills needed to excel in your role as a Journeyman Electrician.

Welcome to the Journey: What This Book Aims to Achieve

This book is like a map for your journey in the world of electricity. It's designed to make the learning process smooth and enjoyable, even if you're completely new to the field. Whether you're starting from scratch or brushing up on your skills, this guide has got you covered.

What's Inside?

1. **Foundations of Electrical Equipment and Installations:** We'll start with the basics, laying a strong foundation for your understanding of electrical bricks. No need to be an expert from the get-go – we'll take it one step at a time.

2. **Tools of the Trade:** You wouldn't set out on a journey without the right gear, would you? We'll explore the tools every Journeyman Electrician needs and how to use them safely and effectively.

3. **Wiring Wonders:** Ever wondered how electricity finds its way to power up your devices? We'll unravel the mysteries of wiring, circuits, and everything in between.

4. **Codes and Regulations:** Just like every journey has rules, the electrical world has codes and regulations. We'll decipher them together, making sure you're on the right path.

5. **Problem-Solving Pitstops:** Every journey has its challenges. We'll tackle common issues and problem-solving techniques, so you can navigate through obstacles like a pro.

6. **Shining a Light on Safety:** Safety is our top priority. We'll discuss safety measures and best practices, ensuring you reach your destination without a hitch.

7. **Putting It All Together:** As we reach the end of our journey, we'll bring everything together. You'll see how the pieces fit, and you'll be ready to tackle real-world electrical tasks confidently.

How to Navigate This Book Efficiently

We know your time is valuable, so let's maximize it. Here are some tips on how to efficiently navigate through this book:

1. **Take it Step by Step:** Don't rush. Each chapter builds on the previous one. Take the time to grasp each concept before moving on to the next.

2. **Hands-On Practice:** Theory is important, but practice makes perfect. Try out what you learn with hands-on exercises. It's the best way to reinforce your understanding.

3. **Use the Glossary:** If you come across a term you don't understand, check the glossary at the end of the book. It's like a dictionary specifically for electrical terms.

4. **Review and Reflect:** At the end of each chapter, take a moment to review what you've learned. Reflecting on the material will help solidify your knowledge.

5. **Ask for Help:** If something doesn't make sense, don't hesitate to ask for help. Whether it's a fellow learner, a mentor, or an online community, there are plenty of resources available.

Remember, becoming a Journeyman Electrician is a journey, not a sprint. Enjoy the process, and before you know it, you'll be confidently navigating the electrical landscape. Let's get started!

Demystifying the Journeyman Electrician Exam

Becoming a Journeyman Electrician is an exciting and rewarding journey that requires dedication, knowledge, and practical skills. As part of the journey, aspiring electricians must pass the Journeyman Electrician Exam, which assesses their understanding of electrical codes, regulations, and safety practices. In this comprehensive guide, we will delve into the details of the exam, including its structure, time limits, passing scores, and the importance of effectively navigating the National Electrical Code (NEC) book.

When preparing for the Journeyman Electrician Exam, it is crucial to familiarize yourself with the types of questions that may appear on the test. The exam typically consists of multiple-choice questions, which require candidates to select the correct answer from a set of options. Some questions may present scenarios or diagrams, testing your ability to apply electrical concepts to real-world situations. It is important to study and understand various topics, including electrical theory, wiring methods, circuit calculations, and safety practices.

The number of questions on the exam can vary depending on the state or jurisdiction. On average, the exam may contain anywhere from 80 to 100 questions. It is essential to manage your time effectively to ensure you have sufficient time to read and answer each question thoroughly. The allotted time to complete the exam usually ranges from 3 hours to 4.5 hours, depending on the state. It is advisable to pace yourself and allocate time for reviewing your answers before submitting the exam.

To pass the Journeyman Electrician Exam, you must achieve a minimum passing score. In most states, the passing grade is set at 70%, meaning you need to answer at least 70% of the questions correctly. However, it is important to note that some states may require a higher passing score, such as 75%. It is crucial to check the specific requirements of your state or jurisdiction to understand the passing criteria.

One aspect that distinguishes the Journeyman Electrician Exam from other exams is the consideration of whether it is an open-book test. In almost all states, the exam allows candidates to reference the NEC book during the test. This means that you can consult the NEC book to find the necessary information and code references to answer the questions. However, it is important to note that even though the NEC book is available for reference, having a solid understanding of the content is crucial for success. Relying solely on the book without a deep comprehension of its content may hinder your ability to locate the right information quickly.

Learning to effectively browse and navigate the NEC book is a crucial skill for success in the Journeyman Electrician Exam. Here are a few examples of how to proceed to find the right information inside the NEC book:

1. **Familiarize yourself with the table of contents:** The NEC book contains a detailed table of contents that outlines the various sections and chapters. Take the time to review and understand the organization of the book. This will help you quickly locate relevant sections when you encounter questions related to specific topics.

2. **Use the index:** The NEC book features an index at the back that allows you to search for specific keywords or terms. When encountering a question that requires a specific code reference, refer to the index to find the relevant page or section that discusses the topic.

3. **Utilize chapter summaries:** Each chapter in the NEC book typically begins with a summary that provides an overview of the topics covered within that chapter. Reading these summaries can give you a general idea of the content and help you identify which chapter to refer to when searching for specific information.

4. **Pay attention to cross-references:** The NEC book often includes cross-references within the text. These references direct you to other sections or articles that provide additional information or clarification on a particular topic. Following these cross-references can lead you to the relevant code sections needed to answer a question accurately.

Remember, while the NEC book is an invaluable resource during the exam, it is essential to develop a solid understanding of the code and its application in real-world scenarios. Use the book as a tool to support your knowledge and problem-solving abilities, rather than relying solely on it. Practice utilizing the NEC book during your exam preparation to become comfortable and efficient in finding the information you need.

Understanding the Exam: Its Role, Importance, and Impact

Role of the Exam: The Journeyman Electrician Exam serves as a crucial checkpoint in your journey. It is designed to evaluate your knowledge, skills, and understanding of electrical concepts and regulations. Successfully passing the exam is often a prerequisite for obtaining a Journeyman Electrician license, a key credential that opens doors to a variety of opportunities in the electrical industry.

Importance of the Exam: The importance of the exam cannot be overstated. It ensures that individuals entering the electrical workforce possess the necessary competence to perform their duties safely and effectively. Additionally, a Journeyman Electrician license is often a requirement for advancing in the field, taking on more responsibilities, and potentially pursuing a Master Electrician certification.

Impact of the Exam: Passing the Journeyman Electrician Exam has a profound impact on your career. It not only validates your knowledge and skills but also enhances your professional credibility. Many employers prefer or require licensed electricians, and clients often seek assurance that the individuals working on their electrical systems are qualified and capable.

Exam Structures and Variations Across States

Exam Structure
Examination questions are designed in a way that assesses the applicant's mastery of the knowledge area. The inclusion of variables in questions ensures that the applicant must apply the appropriate knowledge areas or code rules to reach the correct answer. In instances where a knowledge area has multiple conditions or requirements, the multiple-choice answer selections are crafted to test the applicant's understanding of the subject matter, making it challenging to select the correct answer solely based on key words. Negative-response formats, such as questions asking for what does NOT apply, are used sparingly and serve as a companion format to the multiple-correct answer structure. Plausibility is maintained in incorrect multiple-choice answer selections. Some questions may present extraneous information to evaluate the applicant's ability to discern relevant details.

Throughout the examination, it is assumed, unless specified otherwise, that all questions and related answers operate under a "unity" power factor. Additionally, certain questions address common code violations made by electrical wiring installers. It is emphasized that practical experience should be complemented by quality training to ensure a comprehensive and accurate understanding of electrical code and theory. This underscores the importance of a holistic approach, combining hands-on experience with structured learning to navigate the complexities of electrical work effectively.

Variations Across States
Journeyman electrician exams vary a bit from state to state in the United States. These tests are a crucial step for electricians to get their license and work independently. While the main goal is to ensure that electricians have the knowledge and skills to work safely and effectively, the specific content and format of the exams can differ.

Firstly, let's talk about the commonalities. In general, journeyman electrician exams assess a candidate's understanding of the National Electrical Code (NEC), which is like a rulebook for electrical work. The NEC sets the standards for safe electrical installations, and electricians across the country need to be familiar with it. So, expect questions related to codes, regulations, and safety measures.

However, where the exams differ is in their emphasis on certain topics. Some states might focus more on wiring methods, while others may put greater importance on electrical theory. It depends on the local needs and regulations.

Another factor is the format of the exam. Some states have practical exams where candidates demonstrate their skills by completing electrical tasks. Others stick to written exams with multiple-choice questions. A few states even use a combination of both.

California, for instance, is known for its comprehensive electrical certification process. The exam includes questions on the NEC, but it also covers state-specific regulations. California's emphasis on safety is reflected in its questions about first aid and safety procedures. On the practical side, candidates have to demonstrate their skills in areas like conduit bending.

Texas, on the other hand, has its own unique set of challenges for aspiring journeyman electricians. The Texas Electrical Exam covers wiring methods, equipment and devices, and electrical calculations. Like many states, it includes NEC-related questions, but Texas has its own amendments to the code, so test-takers need to be aware of those as well.

In Florida, the journeyman electrician exam covers topics like grounding and bonding, overcurrent protection, and load calculations. It's essential for candidates to understand the NEC and how to apply it in various situations. Florida, like many other states, also requires a certain amount of work experience or formal education as a prerequisite for taking the exam.

To succeed in these exams, aspiring journeyman electricians need to study not only the NEC but also the specific regulations and amendments in their state. Practical experience is crucial too, as some states require candidates to demonstrate their skills in hands-on scenarios.

While journeyman electrician exams share a common foundation in the NEC, the specific topics, emphasis, and format can vary from state to state. It's important for electricians to thoroughly prepare for their state's exam, understanding both the national and local requirements to ensure they can work safely and effectively in their chosen profession.

Navigating Through State-Specific Licensing and Reciprocity Agreements

To become a certified Journeyman Electrician, candidates must meet specific requirements that can vary significantly from one state to another in the United States. However, there are general guidelines and common criteria that most states follow. Here's an overview to help inform potential candidates about the process and encourage them to check the latest local regulations for the most accurate and relevant information.

Educational and Training Requirements

Most states require candidates to have completed an electrical apprenticeship or vocational training program. These programs typically include both classroom instruction and hands-on training under the supervision of a licensed electrician. The duration and specific requirements can vary, but an apprenticeship usually lasts about 4 to 5 years.

Work Experience

Candidates are generally required to accumulate a certain amount of work experience, often ranging from 4,000 to 8,000 hours of practical, on-the-job training. This experience must usually be verified by a licensed electrician or employer before taking the exam.

Knowledge of the National Electrical Code (NEC)

A thorough understanding of the NEC is crucial, as the exam covers various aspects of electrical installations, including wiring methods, materials, and the proper sizing of electrical components, all of which are governed by the NEC.

Application Process

Before taking the Journeyman Electrician Exam, candidates must typically submit an application to the licensing board or department of their state. This application may require documentation of education, training, work experience, and sometimes a fee.

Examination

The Journeyman Electrician Exam itself tests knowledge of electrical theory, practical skills, and understanding of the NEC and local codes. The format can include multiple-choice questions, calculations, and possibly practical demonstrations, depending on the state.

Additional Requirements

Some states may have additional requirements, such as background checks, continuing education, or specific documentation related to training and experience.

State-Specific Regulations

It's crucial to recognize that each state has its own set of rules and requirements for electrician licensure. For example, the number of hours of work experience required, the specific topics covered on the exam, and the application process can all differ.

Important Note:

Always check the latest and local regulations in your specific state or jurisdiction. State licensing boards or departments typically provide up-to-date information on their websites, including detailed requirements for the Journeyman Electrician Exam, application forms, study guides, and resources for exam preparation.

In summary, while there are common elements in the journey to becoming a Journeyman Electrician across the United States, the specifics can vary widely from one state to another. It's essential for candidates to thoroughly research and adhere to the requirements of their particular state to ensure eligibility to take the exam and achieve certification.

One last word about **Reciprocity agreements** between states: they allow electricians licensed in one state to practice in another without having to undergo the full licensing process again. These agreements are designed to recognize the qualifications and licenses of electricians across state lines, facilitating easier movement and employment opportunities for electricians. However, the specifics of these agreements can vary, with some states having strict criteria for reciprocity or only reciprocating with certain other states. Electricians interested in working in a new state should check for any existing reciprocity agreements and understand the requirements to ensure compliance and eligibility. A table with current Reciprocity agreements can be found at the end of this book.

Psychological and Physical Preparation for the Exam

Preparing for an exam involves both psychological and physical aspects to ensure you're in the best possible shape on the big day.

Psychological Preparation:

1. **Set Realistic Goals:** Break down your study material into manageable sections and set achievable goals. This helps prevent feeling overwhelmed.

2. **Create a Study Schedule:** Establish a study routine that suits your natural rhythm. Some people prefer early mornings, while others are more productive in the evenings.

3. **Positive Visualization:** Imagine yourself succeeding. Visualization can boost confidence and create a positive mindset.

4. **Practice Mindfulness:** Take breaks to clear your mind. Techniques like deep breathing or short walks can help manage stress.

5. **Stay Positive:** Focus on your strengths and previous accomplishments. Avoid negative self-talk.

6. **Connect with Peers:** Discussing topics with others can enhance your understanding and provide different perspectives.

7. **Use Memory Aids:** Create flashcards or acronyms to help remember key concepts. Mnemonic devices can be particularly useful.

Physical Preparation:

1. **Adequate Sleep:** Ensure you're well-rested before the exam. Lack of sleep can affect concentration and memory.

2. **Healthy Diet:** Eat nutritious meals to support brain function. Avoid excessive caffeine or sugar, as they can lead to energy crashes.

3. **Regular Exercise:** Physical activity helps reduce stress and increases overall well-being. Even short walks can be beneficial.

4. **Hydration:** Drink enough water to stay hydrated. Dehydration can negatively impact cognitive function.

5. **Breaks During Study:** Take short breaks during study sessions. Prolonged sitting can lead to fatigue and reduced focus.

6. **Pre-Exam Routine:** Develop a routine before the exam day. This can include a healthy meal, light exercise, and reviewing key points without cramming.

7. **Check Exam Logistics:** Know the exam location and requirements beforehand. This reduces stress on the actual day.

By addressing both psychological and physical aspects, you create a holistic approach to exam preparation. Balancing these elements contributes to a more effective and sustainable study routine, helping you perform your best when it matters most.

Strategic Exam Preparation

Strategic exam preparation involves a thoughtful approach to studying that maximizes efficiency and effectiveness.

Crafting a Well-Rounded Study Plan: Time Management, Breaks, and Consistency

Crafting a well-rounded study plan for the Journeyman Electrician exam involves effective time management, strategic breaks, and consistent effort. Here's a simplified guide to help you create a balanced study plan:

- **Assess Your Starting Point:** Begin by understanding your current knowledge and skills related to the exam content. Identify areas of strength and weakness through self-assessment or practice tests.
- **Set Realistic Goals:** Establish clear and achievable short-term and long-term goals for your study sessions. Break down the exam content into manageable topics to tackle over specific timeframes.
- **Create a Study Schedule:** Develop a weekly or monthly study schedule that fits your routine. Allocate dedicated time for studying each day, considering peak concentration periods.
- **Prioritize Topics:** Identify high-priority topics based on the exam's weightage and your personal strengths and weaknesses. Allocate more study time to challenging areas while maintaining a balance.
- **Active Study Techniques:** Engage in active learning methods, such as practicing with sample questions, creating flashcards, and teaching concepts to someone else. Regularly review and reinforce previously covered material.
- **Time Management:** Set specific time blocks for each study session and adhere to them. Use timers to simulate exam conditions and improve your time management skills. Break down study material into focused segments to enhance comprehension.
- **Strategic Breaks:** Schedule short breaks between study sessions to prevent burnout. Use breaks for light physical activity, relaxation, or a change of scenery. Avoid distractions during study sessions and breaks.
- **Consistency is Key:** Maintain consistency in your study routine. Regular, shorter study sessions are often more effective than sporadic, lengthy ones. Establish a dedicated study space to enhance focus and concentration.
- **Review and Adapt:** Periodically review your study plan and adjust it based on your progress and evolving needs. Be open to adapting your strategies if certain study methods prove more effective.

- **Practice with Sample Exams:** Incorporate practice exams into your study plan to familiarize yourself with the exam format. Analyze your performance in practice exams to identify areas for improvement.
- **Seek Support:** Connect with fellow candidates, join study groups, or seek guidance from experienced electricians. Clarify doubts through discussions and utilize external resources effectively.
- **Maintain a Healthy Lifestyle:** Ensure sufficient sleep, a balanced diet, and regular physical activity to support overall well-being. Manage stress through relaxation techniques and maintaining a positive mindset.

Material Procurement: Where and How to Gather Study Resources

Gathering study resources for the Journeyman Electrician exam is crucial for effective preparation. Here are simple steps on where and how to procure study materials:

1. **Textbooks:** Visit local bookstores or online retailers to find textbooks that cover the topics relevant to the Journeyman Electrician exam. Check if there are specific textbooks recommended by your training program or local licensing board.

2. **Online Resources:** Explore reputable websites that offer study materials for electricians. Websites like Khan Academy, Mike Holt Enterprises, or the National Electrical Contractors Association (NECA) can be valuable. Look for forums or online communities where electricians share study tips and recommended resources.

3. **Local Libraries:** Check your local library for electrical code books, reference materials, and study guides. Libraries often have resources that cover a wide range of electrical topics and can be a cost-effective option.

4. **Training Programs and Courses:** Enroll in a formal training program or online course designed for Journeyman Electrician exam preparation. These programs often provide comprehensive study materials, practice exams, and guidance from experienced instructors.

5. **Trade Schools and Community Colleges:** Inquire at trade schools or community colleges offering electrical courses. They may have recommended textbooks or study guides for the exam. Some institutions also have libraries or resource centers dedicated to trade-specific materials.

6. **Professional Organizations:** Contact professional organizations such as the International Brotherhood of Electrical Workers (IBEW) or the Independent Electrical

Contractors (IEC). These organizations may provide access to study materials, workshops, or networking opportunities with experienced professionals.

7. **Online Forums and Discussion Groups:** Join online forums or discussion groups related to electrical work. Platforms like Reddit or specialized forums can be sources of advice on study materials. Ask for recommendations from those who have successfully passed the Journeyman Electrician exam.

8. **Practice Exams:** Seek out practice exams that mimic the format of the Journeyman Electrician exam. Many study guides and online platforms offer practice questions to help you familiarize yourself with the type of questions asked.

9. **Government Resources:** Check with your local or state licensing board for recommended study materials. Some licensing boards provide official publications or guides that outline the specific content covered in the exam.

10. **Networking with Peers:** Connect with fellow electricians or individuals preparing for the same exam. Share resources and gather insights on which study materials were most helpful for others.

11. **Trade Magazines and Publications:** Subscribe to trade magazines and publications that focus on electrical work. These sources often feature articles, case studies, and updates on industry standards that can be beneficial for exam preparation.

Devising a Systematic Approach for Study and Practice

Creating a systematic approach for studying and practicing for the Journeyman Electrician exam involves organizing your efforts in a structured manner. Here's a simplified guide to help you develop a systematic study plan:

1. **Understand Exam Content:** Obtain a detailed outline of the exam content, including the topics and areas to be covered. Identify the specific codes, regulations, and practical skills expected.

2. **Set Clear Goals:** Define your short-term and long-term goals for exam preparation. Break down your goals into manageable tasks to track progress.

3. **Develop a Study Schedule:** Create a weekly or monthly study schedule based on your available time. Allocate specific time blocks for each exam topic, ensuring comprehensive coverage.

4. **Prioritize Topics:** Identify high-priority topics based on their importance in the exam and your proficiency. Allocate more study time to areas where you need improvement.

5. **Active Learning Techniques:** Engage in active learning methods, such as note-taking, summarizing information, and teaching concepts to others. Use flashcards, diagrams, and practical examples to reinforce understanding.

6. **Practice with Sample Questions:** Incorporate practice questions and sample exams into your routine. Analyze your performance to identify weak areas and focus on improving them.

7. **Simulate Exam Conditions:** Practice under timed conditions to simulate the actual exam environment. Familiarize yourself with any tools or reference materials allowed during the exam.

8. **Review and Adjust:** Review your study plan and adjust it based on your progress. Focus on areas where you face challenges and seek additional resources if needed.

9. **Seek Support:** Connect with peers, study groups, or online forums for mutual support and information exchange. Discuss challenging topics with experienced professionals or instructors.

10. **Practical Application:** Apply theoretical knowledge in practical scenarios. Practice tasks like conduit bending, wiring installations, and troubleshooting. Seek hands-on experience to reinforce theoretical concepts.

11. **Regular Assessments:** Conduct self-assessments periodically to gauge your understanding. Use quizzes, flashcards, or mock exams to reinforce learning.

12. **Establish a Consistent Routine:** Maintain a consistent study routine to develop a habit. Ensure you have a dedicated study space that minimizes distractions.

13. **Utilize Multiple Resources:** Gather a variety of study materials, including textbooks, online resources, practice exams, and video tutorials. Use different resources to gain a well-rounded understanding of the exam content.

14. **Time Management:** Manage your study time efficiently. Avoid cramming and distribute your study sessions evenly. Prioritize difficult topics during your peak concentration periods.

15. **Stay Positive:** Maintain a positive mindset. Celebrate small victories and progress. Visualize success and stay motivated throughout your study journey.

Delving into Mental and Physical Preparedness

Achieving success in preparing for the Journeyman Electrician exam requires a holistic approach that addresses both mental and physical readiness.

Aligning Mind and Body: Strategies for Optimal Performance

When it comes to doing your best as an electrician, it's important to make sure your mind and body are working together. Here are some simple tips to help you perform at your best:

1. **Get Enough Sleep:** Make sure you're getting a good night's sleep. It helps your mind stay sharp and your body stay alert.

2. **Eat Well:** Fuel your body with good food. Eating a balanced diet keeps your energy levels up throughout the day.

3. **Stay Hydrated:** Drink enough water. Dehydration can make you feel tired and less focused.

4. **Take Breaks:** Don't forget to take short breaks during your workday. It helps to recharge your mind and prevent burnout.

5. **Exercise Regularly:** Physical activity is great for your body and mind. It can boost your energy and improve your mood.

6. **Manage Stress:** Find healthy ways to cope with stress. Whether it's deep breathing, taking a walk, or talking to a friend, find what works for you.

7. **Plan Your Day:** Set priorities and plan your tasks. It helps you stay organized and reduces stress.

8. **Continuous Learning:** Keep learning about new developments in your field. It keeps your mind engaged and your skills up-to-date.

Remember, taking care of both your mind and body helps you perform your best as a journeyman electrician.

Stress Management, Nutrition, and Wellness in the Lead-Up to the Exam

When you're getting ready for an exam, it's important to manage stress and take care of yourself. Here are some simple tips for stress management, nutrition, and wellness:

Stress Management:

1. **Break it Down:** Instead of looking at the entire study material at once, break it down into smaller, manageable parts. It makes studying less overwhelming.

2. **Take Breaks:** Don't forget to take short breaks during your study sessions. It helps your mind stay fresh.

3. **Deep Breathing:** Practice deep breathing when you feel stressed. It can help calm your nerves.

4. **Positive Thinking:** Focus on the progress you're making rather than worrying about what you haven't covered yet.

Nutrition:

1. **Balanced Diet:** Eat a balanced diet with a mix of fruits, vegetables, whole grains, and proteins. It provides the energy your brain needs.

2. **Hydration:** Drink enough water. Dehydration can affect your concentration and energy levels.

3. **Limit Caffeine and Sugar:** While a bit of caffeine can help, too much can make you jittery. Also, avoid excessive sugary snacks.

4. **Regular Meals:** Don't skip meals. Regular and balanced meals keep your energy levels steady.

Wellness:

1. **Adequate Sleep:** Make sure you're getting enough sleep, especially the night before the exam. It helps consolidate what you've studied.

2. **Light Exercise:** Incorporate light exercise into your routine. It can help reduce stress and improve your mood.

3. **Connect with Others:** Take breaks to talk to friends or family. A little social time can be a good stress reliever.

4. **Mindfulness or Meditation:** Consider incorporating mindfulness or short meditation sessions into your day. It can help clear your mind.

Remember, taking care of your well-being during the lead-up to the exam is crucial.

Incorporating Physical Activity and Relaxation Techniques in Your Routine

Here are some simple ways to add physical activity and relaxation techniques to your routine:

Physical Activity:

1. **Take Short Walks:** If possible, take short walks during breaks. It could be around your workplace or home. It helps to clear your mind.

2. **Use the Stairs:** Opt for stairs instead of the elevator whenever you can. It's a quick way to add a bit of exercise to your day.

3. **Stretch Breaks:** Stretch your body regularly, especially if you have a desk job. Stretching helps to reduce tension and improve flexibility.

4. **Quick Workouts:** Incorporate quick workouts into your routine, even if they're just 10-15 minutes. It could be simple exercises like squats, lunges, or jumping jacks.

Relaxation Techniques:

1. **Deep Breathing:** Take a few minutes to practice deep breathing. Inhale slowly through your nose, hold for a few seconds, and exhale through your mouth. It helps calm your nervous system.

2. **Mindfulness Meditation:** Spend a few minutes in mindfulness meditation. Focus on your breath and let go of any racing thoughts.

3. **Progressive Muscle Relaxation:** Tense and then gradually release each muscle group in your body. It's a great way to release physical tension.

4. **Listen to Calming Music:** Create a playlist of your favorite calming music. Listen to it when you need a break.

5. **Nature Breaks:** Spend some time outdoors. Whether it's a park or just your backyard, being in nature can be refreshing.

Remember, finding a balance between physical activity and relaxation is key to maintaining overall well-being. Incorporate activities that you enjoy and that fit into your daily routine.

PART II: MASTERING THE NATIONAL ELECTRICAL CODE (NEC)

Introduction to The NEC And Its Structure

The National Electrical Code (NEC), also known as NFPA 70, is a comprehensive set of guidelines and standards that regulate the installation and maintenance of electrical systems in the United States. Developed by the National Fire Protection Association (NFPA), the NEC plays a crucial role in ensuring electrical safety in both residential and commercial structures. Its significance lies in its ability to establish uniformity, promote safety, and adapt to technological advancements in the electrical industry.

Moreover, all editions of the NEC book are available online in different formats. However, individuals may choose to purchase a bound version of the NEC for exam preparation. These versions are equipped with features that align with the specific regulations requested during exams. It is worth mentioning that open-book tests are permitted in almost all states, enabling candidates to consult the NEC book during the examination. However, it is crucial to familiarize yourself with the regulations and requirements of your state or jurisdiction regarding the use of reference materials during exams.

Additionally, **the NEC is typically updated every three years** to incorporate modern technologies, address emerging safety concerns, and enhance the overall effectiveness of electrical systems in the electrical industry. However, it is common for many states to rely on older versions of the NEC rather than adopting the most recent edition. Therefore, it is essential to stay informed about the specific requirements of your state and ensure compliance with the applicable NEC edition during your electrical work.

By utilizing the NEC as a valuable resource, both during exam preparation and in professional practice, you will acquire the necessary knowledge and skills to uphold electrical safety standards and excel as a Journeyman Electrician.

Overview and Significance of the NEC

The NEC serves as a comprehensive guide for electricians, contractors, and authorities having jurisdiction over electrical installations. It covers a wide range of topics, including wiring methods, equipment installation, grounding, and protection against electrical hazards.

One of the key strengths of the NEC is its adaptability. It provides a flexible framework that can be applied to diverse electrical installations, from small residential projects to large-scale industrial facilities. The code is not a one-size-fits-all document; rather, it offers guidelines that can be tailored to specific circumstances while maintaining a baseline standard for electrical safety.

Significance of the NEC:

1. **Safety First:** The primary goal of the NEC is to ensure the safety of individuals and property by minimizing the risk of electrical hazards. Adhering to the code's guidelines helps prevent electrical fires, shocks, and other potentially life-threatening incidents.

2. **Standardization:** The NEC establishes a standardized set of rules for electrical installations. This standardization is crucial for maintaining consistency and predictability in the construction and maintenance of electrical systems across the country.

3. **Legal Compliance:** Compliance with the NEC is often a legal requirement. Many jurisdictions adopt the NEC as part of their local building codes, making adherence to its guidelines mandatory for electrical installations. Non-compliance can lead to legal consequences and may impact insurance coverage.

4. **Technological Advancements:** The NEC evolves to accommodate advancements in technology. As new electrical technologies emerge, the code is updated to ensure that installations incorporating these innovations meet safety standards. This adaptability helps the electrical industry keep pace with technological progress.

5. **Education and Training:** The NEC serves as a valuable educational resource for electricians and professionals in the electrical field. It provides a comprehensive guide to best practices, codes, and standards, contributing to the ongoing professional development of individuals in the industry.

6. **Preventing Electrocutions and Fires:** By establishing guidelines for proper wiring methods, equipment selection, and installation practices, the NEC significantly reduces the risk of electrical accidents, electrocutions, and fires. This, in turn, protects both individuals and property.

The NEC is a cornerstone in ensuring the safety and reliability of electrical systems in the United States. Its continued relevance and adaptability make it an indispensable tool for professionals in the electrical industry, contributing to the overall well-being of communities and individuals through the prevention of electrical hazards. Adhering to the NEC not only meets legal requirements but also reflects a commitment to the highest standards of safety in electrical installations.

A Sneak Peek into the Nine Chapters: A Brief Overview

The National Electrical Code (NEC) is a comprehensive set of standards and guidelines that govern the safe installation, maintenance, and use of electrical systems. In this article, we will delve into the various chapters of the NEC to provide a detailed understanding of its scope and applications.

Introduction (Article 90): The introductory article serves as the foundation for the NEC, outlining its purpose and application. It serves as a guide for electrical professionals, ensuring that installations adhere to standardized safety measures.

Chapters 1-4: General Application: These chapters lay down fundamental principles applicable to all electrical installations, setting the groundwork for safe practices.

- **Chapter 1: General** establishes basic requirements that apply universally to all equipment and installations.
- **Chapter 2: Wiring and Protection** addresses the types of circuits and methods for safeguarding against electrical faults.
- **Chapter 3: Wiring Methods and Materials** focuses on the types of equipment used in distributing electricity, ensuring proper materials and techniques are employed.
- **Chapter 4: Equipment for General Use** identifies and regulates common electrical devices such as lights, motors, and receptacles.

Chapters 5-7: Modification and Supplement: These chapters provide additional specifications and modifications to the general rules outlined in Chapters 1-4.

- **Chapter 5: Special Occupancies** deals with unique environments like hazardous locations, public assembly spaces, and carnivals.
- **Chapter 6: Special Equipment** outlines requirements for specific installations such as IT rooms, fire pumps, and wind turbines.
- **Chapter 7: Special Conditions** addresses conditions like emergency power systems, low-voltage situations, and limited-energy conditions.

Chapter 8: Communication Systems: This chapter, although not subject to Chapters 1-7, becomes relevant when mentioned specifically. It pertains to information technology equipment, including television, radio, and telephone installations.

Chapter 9: Tables: Chapter 9 comes into play where referenced by other chapters, providing essential tables with mandatory reference information crucial for compliance.

Annexes: The annexes serve an information-only purpose and are not mandatory. They provide supplementary information that can aid in understanding the NEC, offering insights into various aspects of electrical installations without imposing strict regulations.

How the NEC Is Structured and How to Navigate It

Understanding the structure and referencing system of the National Electrical Code (NEC) is crucial for anyone involved in electrical installations. The NEC is organized into chapters, each containing specific articles, and further subdivided into sections. This hierarchical arrangement allows for precise identification and easy retrieval of information.

Chapter and Article Numbering: Each chapter of the NEC is assigned a number, and articles within those chapters are numbered accordingly. For instance, "Article 100: Definitions" is located in Chapter 1, providing foundational terms and meanings. As another example, "Article 430: Motors, Motor Circuits, and Controllers" is situated in Chapter 4, addressing regulations related to motors. Similarly, "Article 690: Solar Photovoltaic Systems" is housed in Chapter 6, covering guidelines for solar energy installations.

Section Numbering: Within each article, the information is further organized into sections, identified by a decimal number following the article number. For example, Article 250, which focuses on "Grounding and Bonding," contains sections like 250.10, 250.12, 250.20, and so on. The sections provide specific details and requirements within the broader context of the article.

NEC Section Reference: When referencing a specific section of the NEC, the format is (Chapter Number)(Article Number).Section. For instance, the reference "NEC 250.10" points to Chapter 2 (as indicated by the first digit), Article 250 (Grounding and Bonding), and Section 10, which happens to be titled "Protection of Ground Clamps and Fittings."

Importance of NEC Format: Knowledge of the NEC's format is essential for professionals in the electrical industry. When specifications or technical documents cite NEC sections, individuals familiar with the code can quickly pinpoint the relevant information. This familiarity streamlines communication and collaboration among professionals involved in electrical design, installation, and inspection.

Implementation is Key: While understanding the NEC's format is crucial, the ultimate goal is proper implementation. Adhering to the guidelines and regulations outlined in the NEC ensures electrical installations are safe, reliable, and compliant with industry standards. Professionals must not only be well-versed in the NEC's structure but also diligent in applying its principles to achieve high-quality electrical work.

Detailed Exploration of NEC Chapter 1

Understanding NEC Chapter 1

Introduction

Chapter 1 of NEC 2023, titled "General," is founded upon NFPA 70, 2023 edition, which outlines the minimum installation standards for electrical wiring in commercial, residential, and industrial settings. Originally established in 1897 through collaborative efforts from insurance, electrical, architectural, and related sectors, the code serves as a set of prescriptive guidelines for premises wiring systems. However, it is not intended as a design specification or manual for untrained individuals. The primary objective of the code is to practically safeguard individuals and property against hazards stemming from electricity use.

NEC 2023 adopts this code in its entirety, as indicated in 326B.32 Subd. 2 (3) of Chapter 142, effective July 1, 2023. Published by NFPA, the code encompasses various topics, such as Wiring and Protection, Wiring Methods and Materials, Equipment for General Use, and Special Occupancies, addressing diverse electrical installation requirements and safety measures. The overarching goal is to ensure that electrical systems are installed and maintained in a manner that minimizes hazards while promoting safe and efficient installations.

The "Wiring and Protection" section within the code focuses on fundamental aspects of an electrical system, covering service entrance conductors, grounding, branch circuits, and overcurrent protection. It offers guidelines for sizing and installing conductors and equipment to prevent electrical fires and shocks. The "Wiring Methods and Materials" section details requirements for different wiring methods, including conduit, cable assemblies, and flexible cords, along with specifications for boxes, fittings, cable trays, and raceways.

The "Equipment for General Use" section outlines guidelines for installing and using common electrical equipment like switches, receptacles, lighting fixtures, and appliances to prevent electrical shocks and fires. The "Special Occupancies" section introduces additional requirements for specific occupancies such as hazardous locations, healthcare facilities, and places of assembly, addressing unique risks associated with these environments.

Chapter 1: General of NEC 2023 provides a comprehensive set of guidelines crucial for the safe and efficient installation of electrical systems. This resource is indispensable for electricians, contractors, and inspectors, ensuring the safety of electrical installations.

This article encompasses crucial definitions, specifically those essential to the Code, excluding common technical terms present in other codes and standards.

Accessible equipment refers to items that can be reached for "operation, renewal, and inspection." **Accessible wiring** is capable of being "removed or exposed without damaging" a structure or finish. **Readily Accessible** denotes the ability to reach equipment or items without the need for tools (except for keys) or the removal of interfering equipment.

Ampacity represents the maximum current a conductor can carry without surpassing its temperature rating. The Authority Having Jurisdiction (AHJ) holds the responsibility for enforcing the Code or approving installations and equipment.

Bond and Bonding involve the connection or cable/wire and process used to ensure electrical continuity and conductivity. It is important to note that Bonding is distinct from grounding. **Ground** refers to the earth, while **Grounding** signifies the connection to the ground or the connective body extending to the ground.

Consider a typical distribution system, as depicted in Fig. 2. The definitions for the individual portions, although somewhat self-explanatory, are also included in Art. 100. A **branch circuit** comprises the conductors between the final overcurrent device and the outlet(s). A **continuous load** is one where the maximum current is expected to last for 3 hours or more, and continuous duty involves operating at a substantially constant load for an indefinite period.

Electrical circuits are susceptible to overcurrent conditions, necessitating a system designed for selective coordination. This involves localizing an overcurrent condition to the affected circuit or equipment, isolating it as close to the fault as possible and away from the source. For those designing protective systems, it's crucial to understand that **overcurrent** refers to a fault condition exceeding the equipment's range, potentially causing damage. Overcurrents, resulting from short circuits, ground faults, or overloads, can impact poorly designed systems. **Overloads** occur when current is slightly above the maximum, potentially leading to overheating and generally affecting only one circuit or piece of equipment. As shown in Fig. 2, a fault on the lower branch circuit should cause the fixed overcurrent protective device in the panelboard to open, without affecting protection for the feeders in the power supply source.

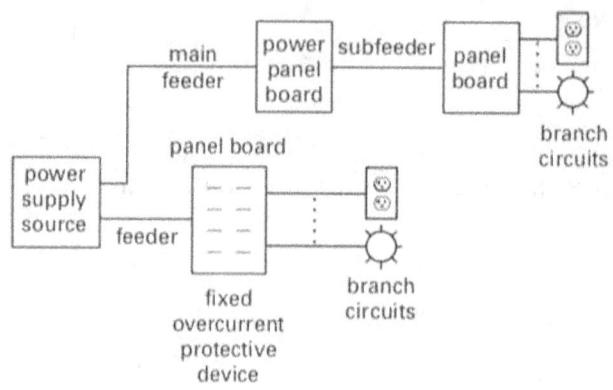

Figure 2: Typical Distribution System

Consult Figure 3 for a detailed explanation of bonding and grounding. The generic terms commonly used by electricians and engineers for grounding wiring may not align with the technical names specified by the NEC. Therefore, understanding these differences is crucial in practical applications.

A **Bonding Conductor or Jumper** is an essential conductor that ensures electrical conductivity between metal parts. In NEC figures, the Bonding Jumper is denoted in yellow. An **Equipment Bonding Jumper** establishes a connection between two or more segments of the **Equipment Grounding Conductor**, which is represented by green wiring. Essentially, when all metal parts lack electrical connection, the bonding jumper ensures continuity to the grounding (green) system. It's important to note that the equipment grounding conductor (green) is not intended to carry current under normal conditions; its presence is for safety in the event of a fault, preventing metal parts from reaching a voltage hazard.

The **grounding electrode**, a conducting object with a direct connection to Earth, is connected through the grounding electrode conductor to the system grounded conductor (neutral—white wire), the equipment grounding conductor (safety ground—green wire), or both.

The Main Bonding Jumper links the grounded circuit (service) conductor (white—commonly known as the "**neutral**") to the equipment grounding conductor (green—commonly known as the "**ground**") or the supply-side bonding jumper, or both. These connections are illustrated in Figure 3.

Quoted terminology represents names commonly used in the field by electricians or those familiar with wiring practices.

In Figure 3, the **panelboard** follows a standard numbering and connection scheme. The black "hot" wire connects to breaker slots #1 and #2, while the red "hot" wire connects to slots #3 and

#4, resulting in a potential of 208 V between them. The potential between the red or black wire and neutral (white) is 120 V.

The "**grounded conductor**," usually the neutral (white wire), exception being a corner grounded delta, is crucial. The diagram illustrates the connection of a single receptacle, detailing the arrangement of wires and the voltage across them.

Breakers in household panel boards provide overcurrent, overload, arc fault circuit interruption (AFCI), and ground fault circuit interruption (GFCI), all topics to be covered in subsequent articles.

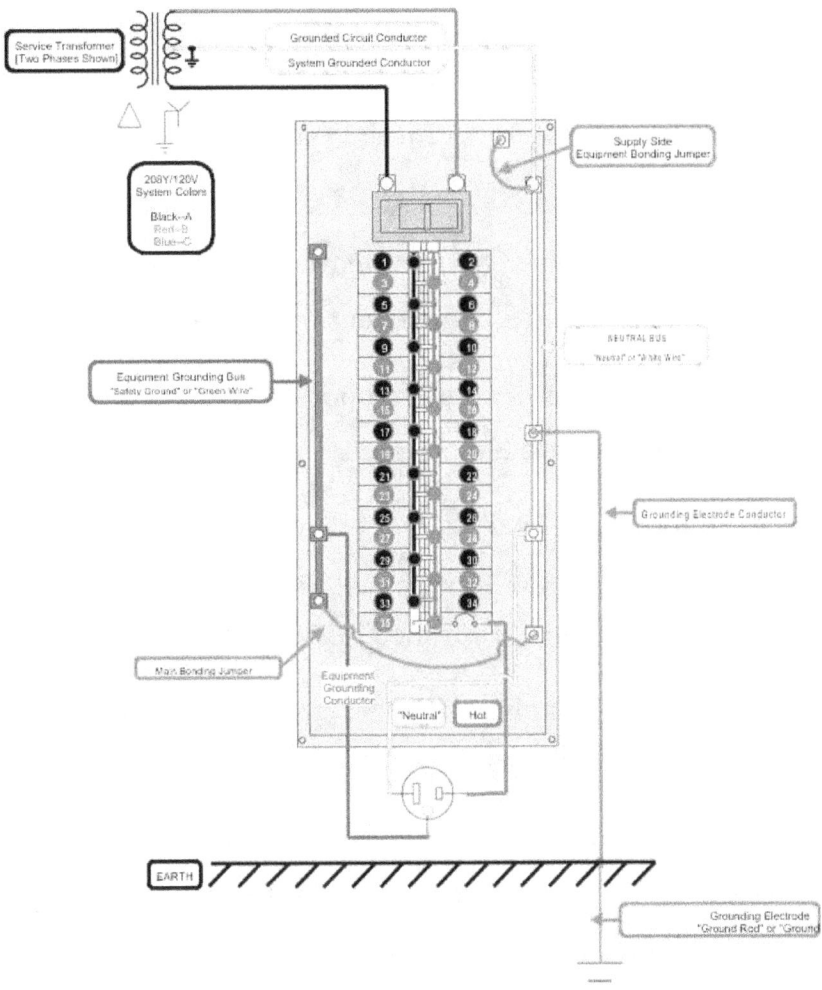

Figure 3: Bonding and Grounding Terminology

The **demand factor** represents the ratio of the system's (or a portion of it) maximum demand to the total connected load. Always below one, it should not be confused with the diversity factor, addressed in IEC 61439, applicable to electrical switchgear designs beyond NEC's scope, particularly in industrial low voltage assemblies.

Differently, the **diversity factor** is the ratio of the combined individual maximum demands of systems to the total connected load. Greater than or equal to one, it considers time, acknowledging that not all maximum loads occur simultaneously. An 80% diversity, for instance, means a device operates at its peak for 80% of its operational time. Residential loads exhibit higher diversity factors, while industrial systems generally have lower values.

In **NEC feeder calculations**, the load is multiplied by the demand factor to reduce overall wiring size. Conversely, for feeders upstream of residential service entrance panels or in industrial facilities with switchgear assemblies, the load is divided by the diversity factor to achieve the same result.

Per Art. 100 definitions, an **electrical datum plane** is a specified distance above water level, permitting electrical equipment installation and connections, accounting for rain and snowfall but excluding manmade or natural disasters. An equipotential plane bonds accessible conductive parts to minimize voltage gradients.

Available fault current is the highest current delivered to a system fault point during a short circuit, constrained by wiring resistance, overcurrent protective devices, and transformer rating.

In the context of conductors, free air denotes an open or ventilated environment allowing heat dissipation and airflow around the conductor.

Island mode refers to an operational state of stand-alone power production equipment or isolated microgrids. Commonly integrating solar equipment and diesel generators, these setups differ from interconnected microgrids as outlined in Art. 705.

Locations are classified as damp (moisture-prone), wet (underground or in direct contact with earth), and dry (not usually damp but potentially temporarily exposed).

A **service** encompasses conductors and equipment linking the serving utility to premises wiring, while a separately derived system, such as generators or transformers, has no direct connection to another source, except incidentally through grounding or metal enclosures.

The **voltage** of a circuit is the greatest root mean square (rms) potential difference between any two conductors. Examples like 208Y/120 V and 480Y/277 V show the first voltages between ungrounded conductors and the second between a conductor and the grounded conductor. Nominal voltage represents a value assigned for convenient designation of a circuit's or system's voltage class, as in 240/120 V, 480/277 V, and 600 V.

Article 110 Requirements for Electrical Installations

This article encompasses the broad requirements pertaining to the installation, utilization, access, and scrutiny of electrical systems (Art. 110.1). Approval for conductors and electrical equipment is imperative (Art. 110.2), typically obtained from the electrical inspection authority, listing laboratories (such as UL—Underwriters Laboratories), or field-based third-party laboratories. Recognized testing laboratories are acknowledged by the Occupational Safety and Health Administration (OSHA) (Art. 110.3(C) Informational Note).

The term "conductors" in the Code specifically denotes copper, aluminum, or copper-clad aluminum. In the absence of explicit specifications, the assumed conductor is copper (Art. 110.5). Conductor sizes are denoted in AWG (American Wire Gauge) or circular mils (cmil or kcmil) (Art. 110.6). AWG is applicable for wire sizes up to 4/0 (referred to as "four aught" or precisely as 0000 AWG), while sizes beyond 4/0 are expressed in circular mils. The sequence of wire sizes ranges from 4/0, 3/0, 2/0, 1/0, 1 AWG through 40 AWG, arranged from largest to smallest. One unit of circular mil equals the area of a circle with a diameter of 0.001 inches. Conversions to more conventional area units are available for reference.

$$A_{cmil} = \left(\frac{d_{inches}}{0.001}\right)^2$$

$$A_{in^2} = 7.854 * 10^{-7} \, A_{cmil}$$

$$A_{cm^2} = 5.067 * 10^{-6} \, A_{cmil}$$

Electrical equipment must be installed with precision and craftsmanship in accordance with Article 110.1, which stipulates a "neat and workmanlike manner." Unfortunately, numerous discrepancies encountered during inspections often do not meet this standard. Though somewhat subjective, guidelines for good workmanship on electrical construction can be found in ANSI/NECA 1-2015.

For systems with a nominal voltage of 1000 V or less, exposed live parts are subject to height, depth, and width restrictions outlined in Article 110.26. The clearance depth is contingent on how the live parts are exposed, detailed in Table 110.26(A)(1). The minimum width must be either the equipment width or 30 inches, whichever is greater, ensuring a 90-degree opening of hinged panels per Article 110.26(A)(2). The clearance height is specified as 2.0 m (6 ½ ft) or the height of the equipment, whichever is greater, as per Article 110.26(A)(3).

Subsequent to this point, specific "Chapter and Part" designations will generally not be employed. While beneficial for quick reference in the table of contents, individual article and section numbers will direct users straight to the required information.

It's noteworthy that NECA stands for the National Electrical Contractors Association.

Key Takeaways for Exam Preparation
The NEC 2023, Chapter 1: General, is a set of rules for electrical wiring in various buildings. It prioritizes safety and is effective from July 1, 2023. It covers topics like Wiring and Protection, Wiring Methods, Equipment for General Use, and Special Occupancies. Electricians, contractors, and inspectors should use it as a guide for safe installations.

In the "Wiring and Protection" section, guidelines for proper sizing and installation of conductors and equipment are given to prevent electrical fires and shocks. The "Wiring Methods and Materials" section details requirements for different types of wiring. "Equipment for General Use" covers safe installation of switches, receptacles, lighting fixtures, and appliances. The "Special Occupancies" section has additional rules for specific places.

Article 100 provides essential definitions. "Accessible equipment" can be reached for operation, "ampacity" is the maximum current a conductor can carry, and "bonding" ensures electrical continuity. Understanding terms like "overcurrent," "grounding," and "continuous load" is crucial for system design. Figures 2 and 3 illustrate bonding and grounding concepts.

The diversity factor and demand factor affect load calculations. Diversity factor considers the time factor, while demand factor is time independent. Feeder calculations use these factors to determine wiring sizes. Electrical datum plane, equipotential plane, and available fault current are essential concepts.

Article 110 outlines general requirements. Conductors and electrical equipment must be approved, and installations should be neat and workmanlike. Exposed live parts have height, depth, and width restrictions. Compliance with ANSI/NECA 1-2015 ensures good workmanship.

Remember, the NEC 2023 is a critical resource for anyone involved in electrical installations, and understanding key definitions and concepts is vital for exam success.

Question Walkthroughs

Sample Questions

1. What is the purpose of Article 110 in the NEC?

   A) Wiring Methods

   B) General requirements for installation, use, and approval

   C) Special Occupancies

   D) Equipment for General Use

2. What are the acceptable conductor materials according to the NEC?

   A) Copper only

   B) Aluminum only

   C) Copper, aluminum, or copper-clad aluminum

   D) Silver only

3. What is the minimum clearance height for exposed live parts in systems of 1000 V nominal or less?

   A) 1 meter

   B) 2.5 feet

   C) 6 ½ feet

   D) 10 feet

Answer and Explanation

1. **Answer: B)** General requirements for installation, use, and approval

Article 110 outlines general requirements for installation, use, access, and approval of electrical systems.

2. **Answer: C)** Copper, aluminum, or copper-clad aluminum

The Code refers to copper, aluminum, or copper-clad aluminum for conductors.

3. **Answer: C)** 6 ½ feet

The clearance height for exposed live parts is 6 ½ feet in systems of 1000 V nominal or less.

Self-Assessment Questions

Practice Questions

1. What is the primary purpose of the NEC 2023, Chapter 1: General?

   A) Design specification

   B) Instruction manual

   C) Safeguarding persons and property

   D) Insurance document

2. When does the 2023 NEC become effective in Minnesota?

   A) January 1, 2023

   B) July 1, 2023

   C) December 31, 2022

   D) October 1, 2023

3. What does the "Wiring and Protection" section of the code cover?

   A) Types of wiring methods

   B) Service entrance conductors

   C) Equipment for general use

   D) Special occupancies

4. Which section provides requirements for specific occupancies like healthcare facilities?

   A) Wiring and Protection

   B) Wiring Methods and Materials

   C) Equipment for General Use

   D) Special Occupancies

5. What is the purpose of a bonding jumper in electrical systems?

   A) To carry current under normal conditions

   B) To ensure electrical continuity

   C) To provide grounding

D) To interrupt the circuit

6. What is ampacity in electrical terms?

   A) Maximum current a conductor can carry without exceeding its temperature rating

   B) Resistance of a conductor

   C) Voltage drop in a circuit

   D) Length of a conductor

7. What does the diversity factor account for in electrical systems?

   A) Time factor

   B) Voltage drop

   C) Conductor size

   D) Resistance

8. What does the term "demand factor" relate to in electrical calculations?

   A) Conductor size

   B) Time factor

   C) Voltage drop

   D) Resistance

9. What is the purpose of the Main Bonding Jumper in an electrical system?

   A) Connection between grounded circuit conductor and equipment grounding conductor

   B) To carry current under normal conditions

   C) Interrupting the circuit

   D) Voltage regulation

10. What is the primary purpose of the "Equipment for General Use" section in the NEC?

   A) Sizing conductors

   B) Safe installation and use of common electrical equipment

   C) Grounding requirements

   D) Special occupancy regulations

11. What is the significance of the available fault current in a system?

    A) Maximum current a conductor can carry

    B) Largest current delivered to the fault point during a short circuit

    C) Conductor resistance

    D) Voltage drop

12. Which organization publishes the NEC, and is it adopted in Minnesota without amendment?

    A) IEEE

    B) NFPA

    C) OSHA

    D) UL

Answers

1. **Answer: C)** Safeguarding persons and property

The primary purpose of the code is the practical safeguarding of persons and property from hazards arising from the use of electricity.

2. **Answer: B)** July 1, 2023

The 2023 NEC is effective in Minnesota from July 1, 2023.

3. **Answer: B)** Service entrance conductors

The "Wiring and Protection" section covers fundamental aspects, including service entrance conductors, grounding, branch circuits, and overcurrent protection.

4. **Answer: D)** Special Occupancies

The "Special Occupancies" section provides additional requirements for specific types of occupancies, such as healthcare facilities.

5. **Answer: B)** To ensure electrical continuity

A bonding jumper is a reliable conductor necessary to ensure electrical conductivity between metal parts.

6. **Answer: A)** Maximum current a conductor can carry without exceeding its temperature rating

Ampacity is the maximum current a conductor can carry without exceeding its temperature rating.

7. **Answer: A)** Time factor

Diversity factor accounts for time, meaning not all the maximum loads will occur at the same time.

8. **Answer: B)** Time factor

Demand factor is time-independent and affects load calculations.

9. **Answer: A)** Connection between grounded circuit conductor and equipment grounding conductor

The Main Bonding Jumper connects the grounded circuit conductor (neutral) and the equipment grounding conductor (ground).

10. **Answer: B)** Safe installation and use of common electrical equipment

The "Equipment for General Use" section covers the installation and use of common electrical equipment.

11. **Answer: B)** Largest current delivered to the fault point during a short circuit

Available fault current is the largest amount of current delivered to the fault point in a system during a short circuit.

12. **Answer: B)** NFPA

The NEC is published by the National Fire Protection Association (NFPA) and is adopted in Minnesota without amendment.

Detailed Exploration of NEC Chapter 2

Understanding NEC Chapter 2

Introduction

Chapter 2 of the NEC 2023, titled "Wiring and Protection," serves as an extensive manual outlining the essential criteria for installing electrical wiring in commercial, residential, and industrial settings. Grounded in the National Fire Protection Association (NFPA) 70, 2023, this chapter is designed to ensure safe and efficient electrical installations.

The chapter commences with a comprehensive exploration of the utilization and identification of grounded conductors. It furnishes guidelines for recognizing terminals, grounding conductors in premises wiring systems, and identifying grounded conductors. This segment plays a pivotal role in guaranteeing the proper grounding of all electrical systems, thereby mitigating the risk of electrical shock and enhancing overall safety.

Subsequently, the chapter delves into the specifics of branch circuits, detailing general provisions, ratings, and required outlets. Branch circuits, extending from the final overcurrent device safeguarding a circuit to the outlets, are addressed to ensure their proper installation and rating for intended use.

The chapter also encompasses feeders, defined as circuit conductors between service equipment, the source of a separately derived system, or another power supply source, and the final branch-circuit overcurrent device. Emphasizing the critical nature of proper installation and maintenance, this section underscores their importance for the safe and efficient operation of electrical systems.

Furthermore, the chapter tackles branch circuit, feeder, and service load calculations. Offering guidelines for accurately calculating loads on these components, the section aims to prevent overloading, which can lead to electrical fires and equipment damage.

A dedicated section on outside branch circuits and feeders is included, outlining additional requirements for circuits and feeders situated outside of buildings or structures. Recognizing their exposure to environmental hazards, this portion mandates extra protections to ensure their resilience.

Chapter 2 of the NEC 2023 provides a comprehensive guide for the secure and effective installation of electrical wiring and protection systems. Addressing various aspects, from the identification of grounded conductors to load calculations, the chapter upholds the highest standards of safety and efficiency for all electrical installations.

Core Concepts and Theories
While this section provides a brief overview of wiring and protection, it primarily focuses on key aspects relevant to Professional Engineers (PE). Specific details pertaining to certain installations, such as motors, standby power, solar setups, and communication systems, will be thoroughly addressed in the respective segments of the course.

Article 200: Use and Identification of Grounded Conductors outlines the identification criteria for insulated grounded conductors with a size of AWG 6 or smaller (Art. 200.6(A)). The following methods are prescribed for identification:

- Continuous white outer finish

- Continuous gray outer finish

- Three continuous white or gray stripes on insulation other than green

- Colored tracer threads in the braid indicating the source of manufacture

- Mineral-insulated, metal-sheathed cable (Type MI) must bear distinctive markings at its terminations

- Fixture wires are to be identified by one or more continuous stripes (Art. 402.8; also see 400.22(A)-(E))

- Aerial cables may comply with the above or employ a ridge on the exterior for identification.

For insulated grounded conductors with a size of AWG 4 or larger, the first three identification methods apply, or white or gray markings on terminations can be used (Art. 200.6(B)).

Different systems' grounded conductors within the same raceway or cable are identified as described above, but distinct from one another (Art. 200.6(D)). For instance, the neutral in a 480Y/277 V system might be gray, while the neutral for a 208Y/120 V system could be white.

Receptacles, plugs, and connectors with terminal connections for the grounded conductor (neutral) must have white coloring, be marked "W" or "white," or be silver (Art. 200.10(B)). The other terminal (hot) should be different, often in brass.

The phrase "What a PE should know" is derived from NCEES guidance for individuals preparing for the Professional Engineer (PE) exam. It emphasizes the importance of certain foundational knowledge that engineers should possess, enabling electricians to navigate complex conduit or raceway systems and correctly identify the neutral for the system at hand.

Notably, a fixture is a stationary piece of equipment, while cord- and plug-connected wiring offers greater mobility.

Article 210 Branch Circuits

Part I General Provisions

Article 210 addresses branch circuits, and it is essential to initiate the exploration at Table 210.3 for information relevant to specific types of branch circuits, regardless of their intended use.

In the case of a branch circuit supplying DC loads, specific markings are required. The positive polarity should be identified with a red outer finish, a red stripe, a plus sign (+), or the words "POSITIVE" or "POS," with markings occurring at least every 24 inches. Conversely, the negative polarity should bear a black outer finish, a black stripe, a minus sign (−), or the words "NEGATIVE" or "NEG," also appearing at least every 24 inches (Art. 210.5(C)(2)(a) and (b)).

A Ground-Fault Circuit Interrupter (GFCI) serves the purpose of deenergizing a device within a defined timeframe when the current to ground surpasses a predetermined threshold. This threshold is set lower than what is necessary to activate the overcurrent protective device of the supply circuit (refer to ground-fault protection of equipment in Art. 100). The device is visually depicted in Figure 4.

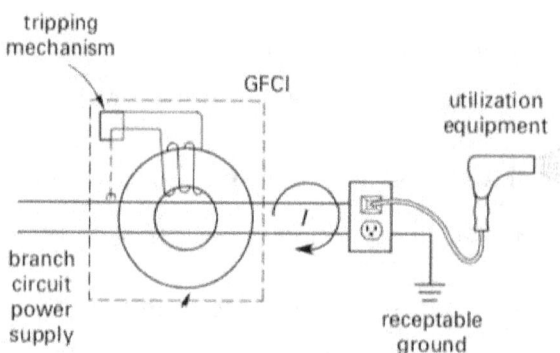

Figure 4: Ground-Fault Circuit Interrupter (GFCI)

GFCIs must be installed in designated locations as outlined in sections 210.8(A)-(F). These locations include bathrooms, garages, outdoor areas, crawl spaces, basements, kitchens, near sinks, boathouses, bathtubs or shower stalls (if within 1.8 m or 6 ft of the outside edge of the tub or stall), laundry areas, and rooftops. Article 422.5 specifies certain appliances that require GFCI protection, such as automotive vacuum machines, drinking water coolers and fill stations, cord- and plug-connected spray washing machines, tire inflation machines, vending machines, sump pumps, and dishwashers. Additional requirements for GFCI are detailed in various sections of the code. A comprehensive list can be found in the Handbook Commentary Table 210.1, and Figure 5 provides a visual reference for a GFCI receptacle.

Figure 5: GFCI 20A Receptacle [Test Black / Reset Gray]

The determination of the necessary branch circuits is typically based on the calculated loads outlined in Article 220.10. Nevertheless, specific branch circuits are mandated according to Article 210.11. These include two 20-ampere small appliance branch circuits, at least one 20-ampere circuit designated for a laundry receptacle, one or more 20-ampere circuits for bathroom receptacles, and a minimum of one 20-ampere circuit for garage receptacles.

A relatively recent addition to the requirements, implemented in 2020, is the necessity for Arc Fault Circuit Interrupter (AFCI) protection on all 120-volt single-phase 15- and 20-ampere branch circuits (refer to Article 210.12). It's important to note that AFCI protection is exempted for branch circuits supplying a fire alarm system, as specified in Article 210.12(B)(1) Exception.

According to Article 210.15, Ground-Fault Circuit-Interrupter (GFCI) protection for personnel or equipment and Arc-Fault Circuit-Interrupter (AFCI) protection cannot undergo reconditioning.

Article 210

Part II Branch Circuit Ratings

The determination of a branch circuit's "rating" relies on the maximum allowable ampere rating or setting of the "overcurrent device," not on the conductor ampacity according to 210.18. Common ratings include 15, 20, 30, 40, and 50 amperes.

For Branch Circuits under 600 V, the minimum ampacity and conductor size information can be found in Art. 210.19, serving as the initial reference for research. Section 210.19(A)(1) directs readers to Art. 310.14, leading to the appropriate ampacity tables. The Informational Note offers advice on voltage drops, recommending a 3% drop for branch circuits to the farthest outlet and a 5% total drop, including feeders. While technically advisory, this note is widely adhered to.

Table 210.21(B)(2) provides the maximum loading for cord-and-plug-connected devices. For instance, a 15 A receptacle should not exceed a load of 12 A. Notably, this represents 80% of the potential loading, a standard design criterion. The 80% (0.8) value aligns with the 125% loading calculation for overload protection, as 100% divided by 80% equals 125%, or 1/0.8 = 1.25. Art. 210.23 details permissible loads for different branch circuit sizes, with a summary in Table 210.24.

Article 210

Part III Required Outlets

Receptacles must be installed within 1.8 m (6 ft) of the intended appliance per Art. 210.50(C). Additionally, general receptacles should not be spaced more than 1.8 m (6 ft) measured horizontally along the floor line of any wall space per Art. 210.52(A)(1) with "wall space" defined as any space 600 mm (2 ft) or more in unbroken floor line per 210.52(A)(2)(1). On countertops, no point along the wall line shall be more than 600 mm (2 ft) from a receptacle per 210.52(C)(1).

Article 215 Feeders

Article 215 covers feeder requirements. And, similar to branch circuits, the minimum conductor size shall have an ampacity not less than 100% of the non-continuous load and 125% of the continuous load. (This is directly related to the 80% design loading mentioned earlier.) This sizing is after correction factors have been applied. Of importance, though the NEC my require the use of the 60°C column ampacity for loading and overload protection, derating can be done on a higher ampacity from the 75°C or 90°C columns if the wire is so rated.

Article 220 Branch-Circuit, Feeder, & Service Load Calculations

Guidance and values for various loads and appliances used in dwellings and non-dwelling units are explained and provided in 220.10. Lighting loads are calculated based on the area covered, that is, the square meters of the units per Table 220.42(A). Demand factors, based on the total load, are used to realistically calculate loading, Table 220.45. Receptacles are included in the lighting load for dwelling units but may be calculated using the demand factors in Table 220.45

for non-dwelling units. Washers, dryers, ovens, and motors are all mentioned but point to other articles for the values. But starting in Art. 220 one will not miss any of the required calculations.

And often used value for receptacles include 180 VA for single, duplex, and triplex receptacles and 360 VA for quad receptacles (90 VA per receptacle). Considering these restrictions, one can put a maximum of 10 receptacles on a 15 A branch circuit and 13 on a 20 A branch circuit.

Demand factors, where allowed, are used to lower the overall load requirements. See Table 1 for a summary of demand factor tables.

Topic	Table	Notes
Lighting Loads	220.42	Based on Total Loading
Receptacles	220.41	Receptacle Loading is added to Lighting Load. No Additional Load Calculations are Required. Use Table 220.42(A)
Dryers	220.54	Based on Number of Dryers
Cooking Appliances	220.55	Based on Power Rating
Kitchen Equipment	220.56	Commercial Equipment
Farm Loads	220.102	Non-Dwelling Unit Loads
Farm Load Total	220.103	Includes Dwelling Loads
Health Care Facilities	220.110(1) & (2)	Based on Category of Spaces
Marinas	220.120	Based on Number of Receptacles

Table 1: Demand Factor Locations

Article 220

Part IV Optional Feeder and Service Calculations

The part refers to calculations for single dwelling units with an ampacity of 100 A or greater.

Article 220

Part V Farm Load Calculations

This part contains calculation guidance for farm buildings. The loads for dwelling units are generally calculated in Parts III or IV of Article 220 and then the farm loads are added to this preliminary total.

Article 225 Outside Branch Circuits and Feeders

An outside branch circuit or feeder is one between buildings, structures, or poles. As is often the case, a table is used to point one to the proper article for a particular application. In this case Table 225.3 points to articles covering everything from irrigation systems, floating buildings, marinas, mobile homes and solar photovoltaic systems.

Conductors for overhead spans of 15 m (50 ft), at 1000 V or less, must be at least 10 AWG copper or 8 AWG aluminum. A longer span requires a minimum 8 AWG copper or 6 AWG aluminum *unless supported by a messenger wire.*

The clearance of outside wires over various objects is found in Art. 225.18.

Article 230 Services

list the various parts within the article and clearly shows the terminology from the serving utility to the branch circuit and should be the first place to look when determining services requirements. The components are summarized in Table 2.

General	Part I
Overhead Service Conductors	Part II
Underground Service Conductors	Part III
Service Entrance Conductors	Part IV
Service Equipment—General	Part V
Service Equipment—Disconnecting Means	Part VI
Service Equipment—Overcurrent Protection	Part VII
Services Exceeding 1000 V, Nominal	Part VIII

Table 2: Article 230 Services Summary

Article 230

Part III Underground Service Conductors

Due to corrosion in underground environments, underground service conductors must be insulated for the voltage applied per Art. 230.30(A) and conduit and wires must be of approved types per 230.30(B).

Article 240 Overcurrent Protection

240.2 Definitions

A current-limiting overcurrent protective device restricts the current to a significantly lower level compared to what would be achievable in the identical circuit using a solid conductor with similar impedance.

Such devices generally operate within one-half cycle and thus result in a let-through energy less than the rating of the components it protects.

240.4 Protection of Conductors

Table 240.3 is the place to start when trying to determine the ampacity and overcurrent protection provided for certain types of equipment. It lists everything from branch circuits to x-ray equipment, pointing to the correct NEC Article to reference. The general requirements for protection are then listed in 240.4.

Section 240.4 states that conductors, other than the flexible type, are protected in accordance with 310.14, which guides the reader to 310.15, which then guides one to the actual ampacity tables of

310.16+. (The + indicates there are many tables beyond 310.16, with each having separate entry criteria such as location, insulation conditions, and so on.) Sections 310.14 and 310.15 contain guidance on adjustments to be made to the ampacities in 310.16+ and thus come first. Such adjustments include temperature limitations, location limitations (such as in a raceway with other conductors) and more.

Table 310.16 is a great place to start when ampacity is required. It is the most commonly used table. Notes at the table bottom will lead one to the proper adjustments.

Section 240.4(B) allows the next higher standard overcurrent device rating, which is above the ampacity of the conductors) if certain conditions are met. *The standard ratings of overcurrent devices can be found in Table 240.6(A).*

Section 240.4(D) covers maximums for small conductors. These values override the values that might be given in 310.16. The two most common gauges used in household and building work 10 AWG and 12 AWG copper. These limits are worth remembering, especially if one works with the NEC often, with 12 AWG limited to 20 A and 10 AWG limited to 30 A.

240.5 Protection of Flexible Cords, Flexible Cables, & Fixture Wires

Having just mentioned the limits for conductors from 240.4(D), this section allows smaller wires to be utilized as "tap" conductors to be attached to circuits protection at values above their ampacities. For example, A 14 AWG wire that is limited to 15 A by 240.4(D)(3) is allowed to be connected to a circuit protected at 20 A or 30 A per 250.5(B)(2)(2) and 250.5(B)(2)(3).

240.12 Electrical System Coordination

Coordination of trip devices is used when an orderly shutdown of a system is required, or when it is desired to maintain the unaffected parts of the system in operation. The main premise of coordination is summarized as opening the breaker closest to the fault first and furthest from the fault last.

240.21 Location in Circuit

Multiple requirements here for taps, according to lengths and locations. One to remember is that a transformers overcurrent protection can be located on the primary side only if the conditions of 240.21(C)(1) are met.

Part III Enclosures

240.33 Vertical Position

Enclosures to be mounted vertically. This is primarily to comply with 240.81 that requires "up" to be "on" and "down" to be "off".

Part VII Circuit Breakers

240.83 Marking

Many marking provisions exist. Of note for most PE's would be the interrupting rating, which, if anything other than 5000 A, must be marked on the breaker per 240.83(C).

Article 242 Overvoltage Protection

This entire article is focused on surge protection devices (SPDs). *Use Table 242.3 for requirements for specialized equipment.*

Article 250 Grounding & Bonding

Much confused, a "ground" is a connection to Earth. Bonding is a connection established in such a way as to ensure electrical continuity and conductivity. So, the ground is the actual connection and bonding is the ohmic requirements for that connection. *Start in Figure 250.1 to find the appropriate Part containing the requirements for which you are interested. Use Informational Note Fig. 250.3 for specific types of equipment.*

Refer back to figure 3 for the various grounding conductor terminology.

250.36 High-Impedance Grounded Neutral Systems

Neutrals installed deliberately with a grounding impedance device (resistor) are known as high-impedance grounded systems. They are used to limit ground fault current to a value that does NOT exceed the *capacitive charging current* and thus does not result in tripped breaker. This is used in systems where continuity of power is at a premium.

The capacitive charging current is the current that flows in a transmission line even when no load is connected. For information, this current can be calculated by summing the zero-sequence capacitance or by determining the capacitive reactance of the cabling and using Ohm's law to determine the current flow.

The cable connecting the grounding device must be rated for the current, but in no case shall it be smaller than 8 AWG copper or 6 AWG aluminum or copper-clad aluminum per 250.36(B).

Part III Grounding Electrode System

250.52 Grounding Electrodes

Common grounding electrodes include water pipes—if in contact with the earth for 3 m (10 ft) or more and metal in ground support structures, again with the same contact requirement.[14] Concrete-encased electrodes of zinc or copper can be used but must have contact of at least 6 m (20 ft). Many other types are listed include the very common rod electrodes, which have contact of 2.44 m (8 ft).

The primary goal (requirement) is for the grounding system to have resistance to earth of 25 Ω per 250.53(A)(2) Exception. This requirement for 25 Ω can be found in numerous references.

The grounding electrode conductor that connects the system to the grounding electrode is sized per Table 250.66.

Part V Bonding

Bonding is the process of connecting material to ensure electrical continuity. The NEC tells one how to bond and the size of the connecting cable, Table 250.102(C)(1). It does not provide an ohmic specification for the bond. This varies according to the type of equipment, connection, and purpose. The value varies from 1 mΩ and upward.

Part VI Equipment Grounding and Equipment Grounding Conductors

The title is self-explanatory with this part providing guidance on the grounding requirements for everything from metal enclosures to pipe organs.

Equipment grounding conductors (see Figure 3) are sized to be up to but not larger than the circuit conductors per 250.122(A). A relatively new requirement (in 2020 Code) requires the equipment grounding conductor to be increased proportionally if the circuit conductors are increased per 250.122(B).

Key Takeaways for Exam Preparation
Chapter 2 of the NEC 2023 focuses on "Wiring and Protection," following guidelines from the National Fire Protection Association (NFPA) 70, 2023. It covers crucial aspects like grounded conductors, branch circuits, feeders, load calculations, and outside circuits, emphasizing safety and efficiency.

Grounded Conductors (Article 200): Grounded conductors (neutral wires) AWG 6 or smaller are identified by white or gray outer finish, continuous stripes, or colored tracer threads. For AWG 4 or larger, the identification methods remain, or markings on terminations. Terminals for grounded conductors are white or silver, distinguishing them from hot terminals.

Branch Circuits (Article 210): Branch circuits, detailed in Article 210, include specifications for DC loads, GFCI installation in specific locations, and requirements for various circuits like small appliance, laundry, bathroom, and garage receptacles. AFCI protection is now mandatory for 120V single-phase 15- and 20-ampere branch circuits.

Branch Circuit Ratings (Article 210): Branch circuit ratings depend on the maximum ampere rating of the overcurrent device, with standard ratings like 15, 20, 30, 40, and 50 amperes. Specific loading limits for cord-and-plug connections are provided to ensure safety and prevent overloading.

Required Outlets (Article 210): Receptacles must be installed within 6 feet of intended appliances. General receptacles should not be spaced more than 6 feet along the floor line of walls. Countertop receptacles must be within 2 feet of any point along the wall line.

Feeders and Load Calculations (Articles 215 and 220): Article 215 covers feeder requirements, emphasizing proper conductor sizing for non-continuous and continuous loads. Article 220 guides load calculations for various appliances, providing tables and demand factors for realistic assessments.

Overcurrent Protection (Article 240): Article 240 defines current-limiting overcurrent protective devices and outlines protection requirements for conductors, including adjustments based on factors like temperature and location. The maximum ampacity for small conductors is specified, ensuring safe usage.

Grounding and Bonding (Article 250): Grounding establishes a connection to Earth, while bonding ensures electrical continuity. Guidelines cover grounding electrode systems, bonding requirements, and sizing of grounding electrode conductors. Equipment grounding conductors must not exceed circuit conductors' size.

Question Walkthroughs

Sample Questions

1. According to Article 240.12, what is the main premise of electrical system coordination?

    a. To increase energy efficiency

    b. To minimize energy consumption

    c. To maintain unaffected parts of the system in operation during a fault

    d. To regulate current flow

2. What does Article 240.33 specify regarding enclosures?

    a. Enclosures must be mounted horizontally

    b. Enclosures must be mounted vertically

    c. Enclosures must be painted in specific colors

    d. Enclosures must be made of specific materials

3. What does Article 250.36 address regarding high-impedance grounded neutral systems?

    a. Capacitive charging current

    b. Grounding electrode conductors

    c. Bonding requirements

    d. Overcurrent protection

Answer and Explanation

1. **Answer: c.** To maintain unaffected parts of the system in operation during a fault

Electrical system coordination aims to maintain unaffected parts of the system in operation during a fault.

2. **Answer: b.** Enclosures must be mounted vertically

Enclosures are required to be mounted vertically, complying with the direction of "up" as "on" and "down" as "off."

3. **Answer: a.** Capacitive charging current

Article 250.36 addresses high-impedance grounded neutral systems, which aim to limit ground fault current not exceeding capacitive charging current.

Self-Assessment Questions

Practice Questions

1. What is the purpose of Chapter 2 of the NEC 2023?

    a. To regulate water usage

    b. To outline installation criteria for electrical wiring and protection

    c. To provide guidelines for plumbing systems

    d. To address fire safety in buildings

2. How are insulated grounded conductors size AWG 6 or smaller identified according to Article 200?

    a. Continuous white outer finish

    b. Continuous red outer finish

    c. Continuous black outer finish

    d. Continuous blue outer finish

3. What is the purpose of GFCIs (Ground-Fault Circuit Interrupters)?

    a. To increase electrical current

    b. To deenergize a circuit when ground current exceeds a predetermined value

    c. To protect against lightning strikes

    d. To regulate voltage in circuits

4. According to Article 210, where should GFCIs be installed?

    a. Anywhere in the building

    b. Only in kitchens and bathrooms

    c. In outdoor locations, basements, and kitchens

    d. Exclusively in garages and laundry areas

5. How is the rating of a branch circuit determined according to Article 210?

a. Based on the conductor's color

b. Based on the maximum permitted ampere rating of the overcurrent device

c. Based on the length of the circuit

d. Based on the voltage of the circuit

6. What is the purpose of AFCI protection for branch circuits according to the 2023 Code?

   a. To regulate current flow

   b. To prevent electrical fires

   c. To enhance energy efficiency

   d. To minimize voltage fluctuations

7. What is the maximum loading limit for a 15 A receptacle according to Table 210.21(B)(2)?

   a. 15 A

   b. 12 A

   c. 20 A

   d. 18 A

8. According to Article 215, what is crucial for sizing feeder conductors?

   a. The color of the conductors

   b. The continuous load

   c. The type of insulation

   d. The voltage of the circuit

9. Where should receptacles be installed according to Article 210.52(A)(1)?

   a. Within 3 feet of appliances

   b. Within 6 feet of intended appliances

   c. At least 2 feet from countertops

   d. On every wall regardless of spacing

10. What does Article 225 of the Code cover?

   a. Grounding and bonding

b. Outside branch circuits and feeders

c. Overcurrent protection

d. Feeder requirements

11. What is the primary goal of the grounding electrode system, as per Article 250.52?

a. To regulate electrical current

b. To ensure electrical continuity

c. To prevent voltage fluctuations

d. To limit ground fault current

12. What is the purpose of overvoltage protection according to Article 242?

a. To increase voltage in circuits

b. To protect against lightning strikes

c. To regulate current flow

d. To minimize energy consumption

Answers

1. **Answer: b.** To outline installation criteria for electrical wiring and protection

Chapter 2 focuses on the minimum installation criteria for electrical wiring in various occupancies, ensuring safety and efficiency.

2. **Answer: a.** Continuous white outer finish

Article 200 specifies that insulated grounded conductors size AWG 6 or smaller should have a continuous white outer finish.

3. **Answer: b.** To deenergize a circuit when ground current exceeds a predetermined value

GFCIs deenergize a circuit when the current to ground exceeds a certain value, preventing electrical shocks.

4. **Answer: c.** In outdoor locations, basements, and kitchens

GFCIs are required in specific locations such as bathrooms, kitchens, outdoors, and basements, among others.

5. **Answer: b.** Based on the maximum permitted ampere rating of the overcurrent device

The rating of a branch circuit is determined by the maximum permitted ampere rating of the overcurrent device.

6. **Answer: b.** To prevent electrical fires

AFCI protection is required to prevent electrical fires in 120V single-phase 15- and 20-ampere branch circuits.

7. **Answer: b.** 12 A

The maximum loading limit for a 15 A receptacle is 80% of its capacity, which is 12 A.

8. **Answer: b.** The continuous load

Feeder conductors must be sized based on the non-continuous and continuous load, considering correction factors.

9. **Answer: b.** Within 6 feet of intended appliances

Receptacles must be installed within 6 feet of the intended appliances according to Article 210.52(A)(1).

10. **Answer: b.** Outside branch circuits and feeders

Article 225 specifically covers requirements for outside branch circuits and feeders.

11. **Answer: d.** To limit ground fault current

The grounding electrode system aims to limit ground fault current, ensuring safety.

12. **Answer: b.** To protect against lightning strikes

Article 242 focuses on surge protection devices (SPDs) to protect against overvoltage, especially due to lightning strikes.

Detailed Exploration of NEC Chapter 3

Understanding NEC Chapter 3

Introduction

Chapter 3 in the 2023 edition of the National Electrical Code (NEC), titled "Wiring Methods and Materials," offers detailed instructions for the installation of electrical wiring systems. The content of this chapter is derived from the National Fire Protection Association (NFPA) 70, 2023 edition.

Commencing with Article 300, the chapter outlines general requirements for wiring methods and materials. This includes the installation and removal of electrical conductors, equipment, and raceways, as well as signaling and communications conductors, equipment, and raceways. Additionally, it encompasses optical fiber cables. Article 300 provides guidance on equipment examination, wiring planning, and stipulates the use and expression of measurements.

A notable change in the 2023 code is the introduction of Article 305, addressing the requirements for wiring methods and materials for systems rated over 1,000 volts AC or 1,500 volts DC. As part of the reorganization, Article 300 is now confined to systems rated 1,000V AC nominal or less and 1,500V DC nominal or less.

Section 300.4(E) specifies the necessity for space separation beneath roof decks and pertains to cables, raceways, or boxes. Initially introduced in the 2008 Code, this requirement now exclusively applies to installations beneath metal-corrugated roof decking.

Section 300.7(B) introduces a new informational note referring to NEMA FB 2.40, which provides installation guidelines on expansion and deflection fittings for raceways.

The revision of Section 300.15 clarifies that a box or conduit body is essential for accessing conductor splices, conductor junction points, and termination points. This clarification emphasizes that the rule pertains to junctions and terminations for conductors, not for conduits.

Section 300.25 outlines the requirements for electrical installations in exit enclosures (stair towers). Stair towers, whether freestanding or integrated with a building, are typically robust, fire-resistive structures constructed from concrete block to enhance passive fire protection for occupants during building evacuation.

Chapter 3 of the NEC 2023 furnishes a comprehensive set of guidelines to ensure the safe and efficient installation of electrical wiring and protection systems. Covering diverse topics, it aims to uphold the highest standards of safety and efficiency for all electrical installations.

Core Concepts and Theories
This article covers all general wiring methods and materials unless it points to other articles.

Chapters 5-7 modify some of the requirements.

310.10(G)(1) Conductors in Parallel

Conductors may be connected in parallel to increase current capability economically. But, the wires must the same length, have the same conductor material and circular mil size, have the same insulation, and be terminated in the same manner, all per 310.10(G)(2).

These requirements are meant to ensure equal current sharing between conductors. For those not working in the building industry, some safety regulations allow parallel conductors, but should one conductor fail, the other conductor must carry the full load—this is NOT an NEC requirement.

Conductor ampacities can be determine from table or calculated using engineering supervision. Most use the tables of 310.16. The Table Notes should be read and utilized for correction factors and conditions that differ from the table assumptions.

310.15 Ampacity Tables

Section 310.15 points to Table 310.16, which is truly the correct place to start for ampacity determine. Section 310.15 contains the Ambient Temperature Correction Factors in 310.15(B)(1) and (2) as well as adjustment factors for more than three current-carrying conductors in 310.15(C)(1). Actual ampacities are in 310.16 through 310.21. An important item to note: *even if another section of the NEC requires you to use the 60°C column, if your wire is rated for 75°C or higher, the adjustment factors are applied to that ampacity, not the 60°C ampacity per 110.14(C).*

312.6 Deflection of Conductors

If you are responsible for installation of wiring in panel boards or cable boxes, normal performed by licensed electricians, an in-depth study of 312.6 and the associated tables is recommended.

Much of the rest of Article 300 covers installation and construction specifications as well as types of cables and their uses.

Key Takeaways for Exam Preparation

Chapter 3 of the NEC 2023 focuses on guidelines for installing electrical wiring systems based on NFPA 70, 2023. The chapter starts with Article 300, providing general requirements for wiring methods and materials, including conductors, equipment, raceways, signaling, and communications. A notable change is the introduction of Article 305, covering systems over 1,000 volts AC, 1,500 volts DC. Section 300.4(E) emphasizes space separation under metal-corrugated roof decking. Section 300.7(B) includes a note referring to NEMA FB 2.40 for expansion and deflection fittings. Section 300.15 clarifies the need for boxes or conduit bodies for access to conductor splices and termination points. Section 300.25 addresses electrical installations in exit enclosures (stair towers).

Chapter 3 covers general wiring methods and materials unless specified otherwise in later articles. Chapters 5-7 modify certain requirements. For conductors in parallel (310.10(G)(1)), they must have the same length, conductor material, circular mil size, insulation, and termination manner (310.10(G)(2)). This ensures equal current sharing, and parallel conductors are allowed by safety regulations. Conductors' ampacities can be determined from Table 310.16, considering correction factors and conditions differing from assumptions. Section 310.15 directs to the correct ampacity determination in Table 310.16, with ambient temperature correction factors and adjustments for multiple conductors. Note that if the wire is rated 75°C or higher, adjust based on that rating.

For panel board or cable box wiring (312.6), an in-depth study is recommended. Article 300 further covers installation, construction specifications, and cable types.

Chapter 3 ensures safe electrical installations by addressing general requirements, parallel conductors, and ampacity determination. It emphasizes space separation, provides guidelines for expansion fittings, and clarifies access needs. The core concepts cover general wiring methods, modifications in later chapters, and specific requirements for parallel conductors and ampacity calculations. An understanding of panel board and cable box wiring is crucial. Studying Article 300 in detail is essential for exam preparation to ensure compliance with the NEC 2023.

Question Walkthroughs

Sample Questions

1. What is the purpose of 312.6 in the context of panel board or cable box wiring?

   A) Determine conductor ampacity

   B) Ensure parallel conductors

   C) Specify deflection of conductors

   D) Provide installation guidelines

2. What type of structures are stair towers mentioned in Section 300.25?

   A) Non-fire resistive

   B) Freestanding

   C) Lightweight construction

   D) Metal-corrugated

3. What does the 60°C column in Table 310.16 represent?

   A) Conductor material

   B) Insulation rating

   C) Circular mil size

   D) Ambient temperature

Answer and Explanation

1. **Answer:** C) Specify deflection of conductors

312.6 addresses the deflection of conductors in the context of panel board or cable box wiring.

2. **Answer:** B) Freestanding

Stair towers are often freestanding structures, independently supported from the building.

3. **Answer:** D) Ambient temperature

The 60°C column in Table 310.16 represents ambient temperature, and adjustments are applied if the wire is rated for 75°C or higher.

Self-Assessment Questions

Practice Questions

1. What does Chapter 3 of the NEC 2023 primarily focus on?

   A) Lighting fixtures

   B) Wiring methods and materials

   C) Electrical appliances

   D) Power generators

2. Which organization's standards serve as the basis for Chapter 3 of the NEC 2023?

   A) IEEE

   B) NFPA

   C) IEC

   D) ASHRAE

3. What is the purpose of Article 300 in Chapter 3?

   A) To regulate electrical appliances

   B) To outline general requirements for wiring methods and materials

   C) To address power generation systems

   D) To specify lighting requirements

4. What is the key change introduced in the 2023 code regarding wiring methods and materials?

   A) Introduction of Article 300

   B) Inclusion of Section 300.15

   C) Addition of Article 305 for systems over 1,000 volts AC

   D) Revision of Section 300.7(B)

5. Which section addresses the space separation requirement under metal-corrugated roof decking?

   A) Section 300.15

   B) Section 300.7(B)

C) Section 300.4(E)

D) Article 305

6. What does Section 300.25 focus on?

   A) Ambient temperature correction factors

   B) Electrical installations in exit enclosures

   C) Expansion and deflection fittings

   D) Conductor ampacities

7. What informational note in Section 300.7(B) refers to installation guidelines?

   A) NFPA 70

   B) NEMA FB 2.40

   C) IEEE 1584

   D) IEC 60364

8. In which section does Chapter 3 provide guidance on the examination of equipment and wiring planning?

   A) Section 300.25

   B) Section 300.15

   C) Article 300

   D) Article 305

9. What is the focus of Chapters 5-7 in relation to Chapter 3?

   A) General wiring methods

   B) Modification of requirements

   C) Conductor parallelism

   D) Panel board wiring

10. What is the primary purpose of connecting conductors in parallel, according to 310.10(G)(1)?

    A) Reduce current capability

B) Increase current capability economically

C) Ensure conductor material uniformity

D) Minimize conductor length

11. What does Section 310.15 direct to for ampacity determination?

A) Section 310.16

B) Section 312.6

C) Article 300

D) Chapter 3

12. What must be considered for conductors' ampacity determination according to Section 310.15?

A) Circular mil size

B) Ambient temperature correction factors

C) Termination manner

D) All of the above

Answers

1. **Answer:** B) Wiring methods and materials

Chapter 3 of the NEC 2023 primarily focuses on guidelines for the installation of electrical wiring systems, emphasizing wiring methods and materials.

2. **Answer:** B) NFPA

Chapter 3 of the NEC 2023 is based on the National Fire Protection Association (NFPA) 70 standards.

3. **Answer:** B) To outline general requirements for wiring methods and materials

Article 300 in Chapter 3 outlines the general requirements for wiring methods and materials in electrical installations.

4. **Answer:** C) Addition of Article 305 for systems over 1,000 volts AC

A significant change in the 2023 code is the introduction of new Article 305, covering requirements for systems rated over 1,000 volts AC.

5. **Answer:** C) Section 300.4(E)

Section 300.4(E) addresses the requirement for providing space separation beneath metal-corrugated roof decking.

6. **Answer:** B) Electrical installations in exit enclosures

Section 300.25 addresses the requirements for electrical installations in exit enclosures, such as stair towers.

7. **Answer:** B) NEMA FB 2.40

Section 300.7(B) includes a new informational note referencing NEMA FB 2.40, providing installation guidelines for expansion and deflection fittings.

8. **Answer:** C) Article 300

Article 300 provides guidance on the examination of equipment and wiring planning, among other general requirements.

9. **Answer:** B) Modification of requirements

Chapters 5-7 modify some of the requirements outlined in Chapter 3.

10. **Answer:** B) Increase current capability economically

Conductors may be connected in parallel to increase current capability economically, as per 310.10(G)(1).

11. **Answer:** A) Section 310.16

Section 310.15 directs to Table 310.16 for ampacity determination.

12. **Answer:** D) All of the above

Section 310.15 considers circular mil size, ambient temperature correction factors, and termination manner for conductors' ampacity determination.

Detailed Exploration of NEC Chapter 4

Understanding NEC Chapter 4

Introduction

Chapter 4 in the NEC 2023, titled "Equipment for General Use," furnishes detailed directives for the installation and operation of commonplace electrical equipment. Derived from the National Fire Protection Association (NFPA) 70, 2023, this chapter encompasses various aspects crucial for electrical safety.

Commencing with Article 400, the chapter addresses Flexible Cords and Cables, offering guidelines on the secure use and installation of these components prevalent in portable devices and temporary setups. Subsequently, Article 402 pertains to Fixture Wires, specifically addressing their application in connecting lighting fixtures and similar equipment, encompassing guidelines for sizing, installation, and usage.

Moving forward, Article 404 delves into Switches, outlining safe installation and usage practices for a variety of switches, including general-use, dimmer switches, and others. Article 406 encompasses Receptacles, Cord Connectors, and Attachment Plugs (Caps), providing guidelines for their secure installation and utilization.

Article 408 concentrates on Switchboards, Switchgear, and Panelboards, elucidating safe practices for their installation and operation. Article 409 then focuses on Industrial Control Panels, prescribing guidelines for their secure installation and use. Article 410 pertains to Luminaires, Lampholders, and Lamps, offering directives for their safe installation and operation.

Article 411 tackles Low-Voltage Lighting, presenting guidelines for the secure installation and operation of low-voltage lighting systems. Lastly, Article 422 covers Appliances, providing comprehensive guidelines for the safe installation and utilization of various appliances.

Chapter 4 of the NEC 2023 establishes an exhaustive set of directives ensuring the safe and efficient installation and operation of commonplace electrical equipment. Addressing a wide array of topics, these guidelines guarantee that all electrical installations adhere to the highest standards of safety and efficiency.

Article 400 Flexible Cords and Flexible Cables

400.4 Types

This section lists the various types of cords and cable along with their properties in Table 400.4 that includes circular mil size, voltage, insulation, and more. The markings on each cable needed to enter the table can be found on the cord or cable at intervals of not more than 610 mm (24 in) apart per 400.6.

400.5 Ampacities for Flexible Cords & Flexible Cables

Tables 400.56(A)(1) and (2) lists the associated ampacities by wire gauge size with Table 400.5(A)(3) having the adjustment factors for the number of conductors. Their permitted and prohibited uses are in 400.10(A) and 400.12, respectively.

402 Fixture Wires

Fixture wire types are in Table 402.3 with ampacities in Table 402.5.

Both flexible cords/cables and fixture wires are not "wiring methods" per the NEC. Hence, they are covered in Article 400 instead of Article 300. Flexible cords/cables connect a device to a power source. Fixture wires are inside those devices, for example, the wiring inside a lamp. Neither type is considered part of the branch circuit.

Article 408 Switchboards, Switchgear, and Panelboards

Ungrounded AC systems are sometimes used for reliability purposes and must be marked per 408.3(F)(2). High-Impedance Grounded Neutral AC Systems are used to help operators identify the location of the fault using ground detectors without taking the equipment off-line. Because this results in a higher voltage to ground, such systems are also labeled to indicate the potential hazard per 408.3(F)(3).

Article 410 Luminaires, Lampholders, and Lamps

Luminaires must be mounted 900 mm (3 ft) horizontally and 2.5 m (8 ft) vertically from a bathroom or shower area, 410.10(D)(1). If located within this zone they must be listed for wet locations, 410.10(D)(2).

Part XV Decorative Lighting and Similar Accessories

Part XVI Special Provisions for Horticultural Lighting Equipment

Article 422 Appliances

Most of the typical calculations required for appliances was covered in Section 220.

Article 430 Motors, Motor Circuits, and Controllers

Part I General

This is the place to start any search for information, *see Figure 430.1 that details what each part covers and provides a visual aid of those components.* See Table 430.5 for specialty applications such as air-conditioning equipment, cranes, and theaters.

430.6 Ampacity and Motor Rating Determination

Nameplate ampere rating causing quite a bit of confusing in the industry. Section 430.6(A)(1) requires the use of the *horsepower rating* on the motor as the entry information to obtain the ampacity from Table 430.247, 430.248, 430.249, and 430.250. (Except for low speed motors, <1200 rpm, high torque and multispeed motors.) The reason for this is that the tables account for typical efficiencies, power factors, and rpm changes that result in higher currents whereas the nameplate does not. The sentence that confuses states, "Where a motor is marked in amperes, but not horsepower, the horsepower shall be assumed to be that corresponding to the value given in Table 430.247, 430.248, 430.249, and 430.250." What this essentially means is that one is using the nameplate amperage. That is, one looks up the nameplate amperage and determines the horsepower rating from the table, interpolating if required, and uses that value for all calculations.

The exceptions to 430.6(A)(1) are important. Exception #1 points to 430.22(B) and 430.52 for amperage guidance. Exception #2 allows the use of the nameplate amperage for shaded-pole or permanent-split capacitor-type fan or blower motor. Exception #3, another source of confusion, allows the use of the *appliance nameplate amperage* even when both the *amperage and horsepower of the motor* are listed. In simple terms, this is because the appliance contains more than just the motor.

430.7 Marking on Motors and Multimotor Equipment

In 430.7(A)(9) specifies that a motor shall be marked with a Design Letter of A, B, C, or D that relate motor characteristics such as current, slip, locked-rotor and breakdown torque. These letters are defined in ANSI/NEMA MG-1-1993, *Motors and Generators, Part 1, Definitions* and IEEE 100-1996, *Standard Dictionary of Electrical and Electronic Terms.*

These design letters should not be confused with the *locked rotor indicating code letters*, which also have an A,B, C, and D and are listed in Table 430.7(B).

430.22 Single Motor

Motors don't always operate continuously. When they don't, the conductor ampacities may be reduced based on the nameplate amperage and values given in Table 430.22.

430.24 Several Motors or a Motor(s) and Other Load(s)

Generally, branch circuit conductors, supplying multiple loads, for motors are protected at 125% of the full-load current rating of the highest rated motor, sum of the current ratings of all the other motors, 100% of the noncontinuous non-motor load, and 125% of the continuous non-motor load.

430.52 Rating or Setting for Individual Motor Circuit

Up to now, the discussion has centered on overloads. The overcurrent/short-circuit protection can be found directly in Table 430.52. It depends on the type of motor as well as the type of protection. For most households, circuit breakers combine overload and short-circuit protection in one device.

When the size of the short-circuit and ground-fault protective device from table 430.52 does not correspond to a standard rating, the next higher standard rating is allowed, 430.52(C)(1) *Exception No. 1*. See 210.18 and Table 240.6(A).

Part X Adjustable-Speed Drive Systems

Recommend study if dealing with such systems.

Article 480 Storage Batteries

The article itself covers stationary battery installations. It starts with a note listing many IEEE and UL standards that are applicable. These standards are in an "information note" and as such are not part of the requirements of the NEC, but by listing them they clearly should be read and understood. Figure 6 shows some battery terminology.[18]

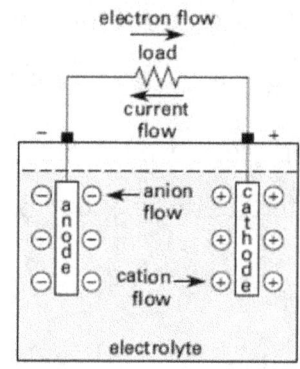

Figure 6: Battery Standard Terminology

480.2 Definitions

Nominal Voltage (Battery or Cell): The value assigned to a cell or battery for the purpose of convenient designation. The operating voltage varies depending on a variety of factors.

Informational Note: Lead-Acid has a nominal cell voltage of 2 V/cell. Alkali systems have a nominal of 1.2 V/cell. Li-ion cells, now in widespread use to due high energy density, have a nominal voltage of 3.6 V/cell to 3.8 V/cell.

480.6 Overcurrent Protection for Prime Movers

The cable running from the battery in your car to the starter has no overcurrent protection. This is because of 480.6 that allows batteries with a voltage of 60 V or less to forego overcurrent protection if they provide power for starting, ignition, or control of prime movers.

480.10 Battery Locations

Battery spaces require ventilation to prevent an explosive mixture from forming. Mechanical ventilation may not be required. Hydrogen disperses easily though. It accumulates at the top of spaces and a means of removal must be installed. Certain batteries called "Valve-Regulated" are considered to be sealed but even during normal operation may emit some hydrogen.[19] During failure of such a battery large amounts of explosive gasses can be released. Only Li-Ion and Nickle Chloride do not require ventilation during normal and abnormal charging conditions.

Key Takeaways for Exam Preparation
Chapter 4 of the NEC 2023 focuses on the "Equipment for General Use" and is based on the National Fire Protection Association (NFPA) 70, 2023. It covers a range of topics from flexible cords and cables to appliances and storage batteries.

In Article 400, "Flexible Cords and Cables," guidelines for the safe use and installation of these components are provided. It includes types, ampacities, and permitted/prohibited uses.

Article 402, "Fixture Wires," outlines guidelines for connecting lighting fixtures, emphasizing proper sizing, installation, and use.

Article 404, "Switches," covers the installation and use of various switches, ensuring safety in their application.

Article 406, "Receptacles, Cord Connectors, & Attachment Plugs," provides guidelines for safe installation & use of these common electrical components.

Article 408, "Switchboards, Switchgear, and Panelboards," outlines safety measures for the installation and use of these systems.

Article 410, "Luminaires, Lampholders, and Lamps," specifies mounting requirements for luminaires and emphasizes wet location listings for those within specific zones.

Article 422, "Appliances," provides guidelines for the safe installation and use of appliances.

Article 430, "Motors, Motor Circuits, and Controllers," covers general information about motors, including ampacity determination, markings on motors, and considerations for single or multiple motors.

Article 480, "Storage Batteries," addresses stationary battery installations, with definitions, overcurrent protection, and considerations for battery locations.

Understanding core concepts like the types of cords and cables, ampacities, and the proper use of fixtures, switches, and receptacles is crucial. Knowledge of motor-related topics, including ampacity determination, motor markings, and considerations for multiple motors, is essential. Additionally, awareness of storage battery definitions, overcurrent protection, and safety measures for battery locations is important.

Question Walkthroughs

1. What does Article 480 address in the NEC 2023?

   A) Motors

   B) Storage Batteries

   C) Luminaires

   D) Appliances

2. According to 480.6, when is overcurrent protection not required for batteries?

   A) For all batteries

   B) For batteries with a voltage of 60 V or less

   C) Only for lead-acid batteries

   D) Only for Li-ion cells

3. What is the purpose of mechanical ventilation in battery spaces, as mentioned in 480.10?

   A) To prevent hydrogen accumulation

   B) To cool the batteries

   C) To increase battery life

   D) To prevent electrical shocks

Answer and Explanation

1. **Answer: B)** Storage Batteries

Article 480 addresses stationary battery installations.

2. **Answer: B)** For batteries with a voltage of 60 V or less

Overcurrent protection is not required for batteries with a voltage of 60 V or less if they provide power for starting, ignition, or control of prime movers (480.6).

3. **Answer: A)** To prevent hydrogen accumulation

Battery spaces require ventilation to prevent an explosive mixture from forming, particularly hydrogen.

Self-Assessment Questions

Practice Questions

1. What does Chapter 4 of the NEC 2023 focus on?

   A) Lighting fixtures

   B) Equipment for General Use

   C) Flexible Cords and Cables

   D) Appliances

2. Which article covers guidelines for the installation of flexible cords and cables?

   A) Article 400

   B) Article 402

   C) Article 404

   D) Article 406

3. What is the purpose of Article 402 in the NEC 2023?

   A) Guidelines for switches

   B) Guidelines for receptacles

   C) Guidelines for fixture wires

   D) Guidelines for motors

4. According to Article 410, how should luminaires be mounted near bathroom or shower areas?

   A) 1 ft horizontally and 3 ft vertically

   B) 2 ft horizontally and 2 ft vertically

   C) 3 ft horizontally and 2.5 ft vertically

   D) 4 ft horizontally and 3 ft vertically

5. In Article 430, what information is required for determining ampacity?

   A) Voltage rating

   B) Horsepower rating

C) Nameplate amperage

D) Motor speed

6. What is the purpose of Design Letters in motor markings (430.7(A)(9))?

    A) Indicate manufacturing date

    B) Provide ampacity information

    C) Relate motor characteristics

    D) Specify voltage requirements

7. When does Exception #3 in 430.6(A)(1) allow the use of appliance nameplate amperage?

    A) For all motor types

    B) For shaded-pole or permanent-split capacitor-type motors

    C) Only for low-speed motors

    D) Only for high-torque motors

8. Why might ungrounded AC systems be used, as mentioned in Article 408?

    A) For energy efficiency

    B) For reliability purposes

    C) For higher voltage output

    D) For cost savings

9. Which article covers guidelines for Low-Voltage Lighting systems in the NEC 2023?

    A) Article 404

    B) Article 409

    C) Article 411

    D) Article 422

10. What is the purpose of Article 406 in the NEC 2023?

    A) Guidelines for switches

    B) Guidelines for receptacles

    C) Guidelines for motors

D) Guidelines for storage batteries

11. Where can information about the types of cords and cables be found in Article 400?

    A) Table 400.5(A)(3)

    B) Table 400.56(A)(1)

    C) Table 400.4

    D) Table 402.5

12. What is the purpose of Article 411 in the NEC 2023?

    A) Guidelines for Low-Voltage Lighting systems

    B) Guidelines for storage batteries

    C) Guidelines for appliances

    D) Guidelines for motors

Answers
1. **Answer: B)** Equipment for General Use

Chapter 4 of the NEC 2023 focuses on "Equipment for General Use."

2. **Answer: A)** Article 400

Article 400 covers guidelines for the installation of flexible cords and cables.

3. **Answer: C)** Guidelines for fixture wires

Article 402 provides guidelines for fixture wires used to connect lighting fixtures.

4. **Answer: C)** 3 ft horizontally and 2.5 ft vertically

Luminaires must be mounted 900 mm (3 ft) horizontally and 2.5 m (8 ft) vertically from a bathroom or shower area (410.10(D)(1)).

5. **Answer: C)** Nameplate amperage

Section 430.6(A)(1) requires the use of the horsepower rating on the motor as the entry information to obtain the ampacity.

6. **Answer: C)** Relate motor characteristics

Design Letters indicate motor characteristics such as current, slip, locked-rotor, and breakdown torque.

7. **Answer: B)** For shaded-pole or permanent-split capacitor-type motors

Exception #3 allows the use of the appliance nameplate amperage for shaded-pole or permanent-split capacitor-type fan or blower motor.

8. **Answer: B)** For reliability purposes

Ungrounded AC systems are sometimes used for reliability purposes (408.3(F)(2)).

9. **Answer: C)** Article 411

Article 411 covers guidelines for Low-Voltage Lighting systems.

10. **Answer: B)** Guidelines for receptacles

Article 406 covers guidelines for the safe installation and use of receptacles.

11. **Answer: C)** Table 400.4

Information about the types of cords and cables, along with their properties, can be found in Table 400.4.

12. **Answer: A)** Guidelines for Low-Voltage Lighting systems

Article 411 provides guidelines for the safe installation and use of Low-Voltage Lighting systems.

Detailed Exploration of NEC Chapter 5

Understanding NEC Chapter 5

Introduction

Chapter 5 of the NEC 2023, titled "Special Occupancies," furnishes detailed directives for electrical installations in specialized settings. Grounded in the National Fire Protection Association (NFPA) 70, 2023, this chapter serves as a crucial reference.

Commencing with Article 500, the chapter addresses Hazardous (Classified) Locations. These are areas where potential fire or explosion risks exist due to the presence of flammable gases, vapors, dust, or fibers. Article 500 offers explicit guidelines for the secure installation of electrical equipment in such locations.

Following suit is Article 501, delving into Class I Locations. These are places where flammable gases or vapors may be present in quantities sufficient to create explosive or ignitable mixtures. The article lays out safety guidelines for the installation of electrical equipment in these specific environments.

Article 502 addresses Class II Locations, characterized by hazards arising from combustible dust. The guidelines provided in this article ensure the secure installation of electrical equipment in areas where such dust poses a risk.

Moving on, Article 503 tackles Class III Locations, which are hazardous due to the presence of easily ignitable fibers or flyings. The guidelines within this article are designed to facilitate the safe installation of electrical equipment in these particular locations.

Article 504 centers on Intrinsically Safe Systems, where all electrical circuits are inherently safe, incapable of releasing sufficient electrical or thermal energy to ignite a hazardous atmospheric mixture. The guidelines set forth in this article contribute to the secure installation of these systems.

Article 505 takes on Zone 0, 1, and 2 Locations—sites where an explosive gas atmosphere is either continuously present or likely to occur during normal operations. This article provides essential guidelines for the safe installation of electrical equipment in these zones.

Chapter 5 of the NEC 2023 delivers a comprehensive framework of guidelines to ensure the safe and efficient installation of electrical systems in special occupancies. Covering diverse topics, these guidelines uphold the highest standards of safety and efficiency for all electrical installations.

Core Concepts and Theories
Article 500 Hazardous (Classified) Locations, Classes I, II, & III. Divisions 1 and 2

Requirements for electrical installations in areas designated as hazardous locations go beyond the standard provisions of the NEC. NEC Art. 500, titled "Hazardous (Classified) Locations," outlines the necessary measures to prevent electrical equipment or systems within these locations from igniting flammable or combustible materials. It's important to note that the NEC itself doesn't classify these locations; instead, classification is delegated to the relevant NFPA standards.

NEC Art. 500 incorporates well-known standards within the electrical engineering field, such as NFPA 497, which provides guidelines for classifying flammable liquids, gases, or vapors, and for hazardous (classified) locations in chemical process installations. VRLA, standing for Valve-Regulated Lead Acid, is also mentioned. The classification depends on the hazardous material present and its usage in a specific facility. Once determined in accordance with the appropriate standard, the corresponding electrical requirements of the NEC are applied.

Additionally, NEC Art. 500 references NFPA 499, a recommended practice for classifying combustible dust and hazardous (classified) locations in chemical process areas. It's worth noting that the terms "hazardous" and "classified" are used interchangeably in the titles of standards and codes.

NEC 500.4 (B), Informational Notes 1-6, provides a list of reference standards, including NFPA 30, the Flammable and Combustible Liquids Code, which is commonly encountered by electrical engineers. Recognizing the increased fire hazard posed by flammable aerosol products, a specific version of this code, NFPA 30B, focusing on the manufacture and storage of aerosol products, was developed.

These hazardous standards encompass information primarily related to classification, fundamental facility requirements, storage of hazardous material, occupancy limits, and operational and maintenance requirements. While these standards influence the overall design of a facility, it's essential to emphasize that electrical engineers should pay attention to the design limitations directly imposed on electrical systems, nearly all of which trace back to the relevant articles in the NEC.

Each standard categorizes requirements based on the characteristics of flammable gases or vapors (Class I), combustible dusts (Class II), or fibers/flyings (Class III) that may be present.

Each class is then subdivided into divisions, primarily based on whether the hazard is present during normal operations (Division 1) or abnormal conditions (Division 2). NEC Art. 500 addresses Classes I, II, and III, encompassing Divisions 1 and 2.

Class I locations can alternatively be divided into zones (rather than divisions). These zones pertain to the presence of ignitable concentrations of flammable gases or vapors: continuously or for extended periods (Zone 0), under normal operating conditions or frequent maintenance (Zone 1), or for a brief period or within enclosed containers (Zone 2). Class II and Class III can also be treated as zones, with dust (Class II) or fibers/flyings (Class III).

Refer to NEC Art. 500 Informational Note for NFPA 497, focusing on liquids, gases, or vapors, and NFPA Standard 499, emphasizing dust. NFPA 30B addresses aerosols, and while not explicitly mentioned in NEC Art. 500.4(B), it is referenced in Informational Note 2 as a follow-up to NFPA 30. None of these standards covers explosives, pyrotechnics, or blasting agents—similarly, the NEC does not. Given their unique characteristics, separate standards unrelated to the NEC govern explosives and blasting agents, as pyrotechnics pose a spontaneous ignition risk in air beyond the scope of NEC requirements.

Distinguishing between flammable and combustible liquids, the former has a flash point below 100°F, while the latter has a flash point at or above 100°F. "Flammable" implies easy ignition, while "combustible" necessitates more robust conditions for ignition. Generally, a combustible material is one prone to burn or explode. The flash point denotes the lowest temperature at which the material's vapors will ignite when exposed to an ignition source.

Detailed hazardous classifications will likely be specified in the design. Focus on NEC 70 requirements for locations with continuous or prolonged hazardous conditions (Zone 20), those under normal operating conditions or frequent maintenance (Zone 21), or those with short-term exposure or enclosed in containers (Zone 22).

Requirements for divisions and zones run parallel, with zone designations guided by the International Electrotechnical Commission (IEC). NEC 505.2 defines zone protection methods. In the zone system, no differentiation exists for fibers/flyings, and material groups are numbered differently. Refer to Table 3 for a comparison of groups in the division system and the zone system.

NEC Art. 501 outlines division requirements for Class I, Art. 502 covers Class II division requirements, and Art. 503 addresses Class III division requirements. The specifications for intrinsically safe systems can be found in Art. 504. Class I zone requirements are detailed in NEC Art. 505, while Art. 506.25 outlines Class II and Class III zone requirements.

To further distinguish hazardous conditions within each class, atmospheric groups are assigned. Each class comprises multiple groupings: Class I includes Groups A, B, C, D, and E, while Class II includes Groups E, F, and G.

For equipment testing, approval, and area classification in Class I Zones 0, 1, and 2, Material Groups IIA, IIB, and IIC are used, arranged in order of severity. Similarly, when Class II (dusts) and Class III (fibers/flyings) locations are categorized by zones instead of divisions, they are divided into three material groups: IIIA, IIIB, and IIIC, again listed in order of severity. These material groups are crucial for testing and rating equipment suitable for specific zones.

Additional guidance for designing intrinsically safe (IS) equipment intended for use around these materials and within specified hazardous divisions or zones can be found in NEC Art. 504.

Division System	Zone System
Groups	Groups
IIC	A and B
IIB	C
IIA	D
I	D

Table 3: Group Equivalency between Division and Zone Classifications

An intrinsically safe (IS) system is one in which the circuit does not develop sufficient energy (measured in mJ) in an arc or spark to cause ignition, nor can an overload condition develop sufficient thermal energy to cause the ignition temperature to be exceeded.

NEC Art. 504, provides the requirements necessary to design an intrinsically safe (IS) piece of equipment for any of the locations mentioned from Art. 500-516.

Hazardous Location Classification: Aerosol Products

NFPA 30B focuses on the hazards of aerosol products. It requires electrical equipment and wiring to meet general NEC requirements.[26] However, when aerosols are installed in areas where flammable liquids and gasses are handled, NEC Arts. 500 and 501 must be met. Limiting sources of ignition, such as lightning, static electricity, arcs and sparks, and stray currents, are a concern. These hazards are minimized by complying with NEC requirements during installation. For example, lightning protection should be provided using rods, grounding, and surge protectors.[29] When a facility has a flammable propellant and charging pump room, the electrical equipment and wiring in the room must meet the requirements of a Class I, Division 1 location or Class I, Zone 1 location according to NEC Art. 500, 501, 504, and 505.

Hazardous Location Classification: Liquids, Gasses, and Vapors

NFPA 497 focuses on the hazards of liquids, gasses, and vapors in chemical process areas. Combustible materials are divided into classes. Class I division requirements apply to specific materials with a flash point of less than 100 F and is further divided into four groups. Group A is acetylene. Group B is a gas, liquid, or vapor that may burn or explode with a minimum experimental safe gap (MESG) of less than or equal to 0.45mm or a minimum igniting current ratio (MIC ratio) of less than or equal to 0.40—typically, hydrogen. Group C and D have higher MESG or MIC ratios than Group B.

NFPA 30B 4.3.1.1

NFPA 30B 4.8.2 and 8.3. Only those sources directly concerning electrical engineering are listed in this manual.

See the *NEC Handbook* index for a list of the numerous articles covering these topics.

NFPA 30B 5.3

Class I zone requirements apply to locations within a facility where combustible material mixes with air. Zones are divided into Groups IIA, IIB, and IIC as determined by the MESG or MIC ratio. Tables listing various chemicals and their classifications are provided.

The classifications are further delineated in NFPA 497, Chap. 5 as Class I, Division 1 or 2, and Class I, Zone 0, 1 or 2. The classifications are dependent upon the hazardous material, whether it is present in the atmosphere frequently or not, the ventilation system, and the potential impact to electrical systems. These zones are then used to determine the applicable articles in the NEC.

The items impacting electrical design and installation from NFPA 497 are incorporated into the NEC. Therefore, once the classification is known, or provided in design documentation, the applicable sections of the NEC are then applied to the electrical design—NEC Art. 500, 501, 502, 503, and 505.

Hazardous Location Classification: Combustible Dust

NFPA 499 focuses on combustible dust in chemical process areas. It divides combustible dusts, considered Class II, into Groups E, F, and G depending upon the presence of combustible dust in atmosphere. Class III, Groups IIIA, IIIB, and IIIC are material groups where combustible dust is present, being handled, manufactured, or stored.

- Dust ignition-proof electrical equipment, pressurized electrical equipment, and intrinsically safe (IS) equipment can be used in both Division I and Division 2 locations. Other dust tight equipment enclosures as specified in NFPA 70 (the NEC) may be used in Division 2 locations. However, electrical wiring or equipment approved for Class I, Division 1 (normal operations) may not be suitable in a Class II location. Protection for an electrical breakdown resulting in an ignition is not provided in the standard because the possibility of changes in the breakdown voltage and the release of combustible material occurring simultaneously is considered remote. Where both flammable gasses or vapors and combustible dust are present, the design must comply with both Class I and Class II conditions. Combustible dusts are divided into three groups in the Division system (Groups E, F, and G) and three groups in the Zones system (Groups IIIC, IIIB, IIIA).

The various combustible dusts along with their properties are listed in NFPA 497 Table 5.2.1 using their CAS number (Chemical Abstract number). Ignition temperatures are provided to place the dust into the appropriate material group. Once the materials are known, the process for properly classifying an area is given in NFPA 497 Sections 6.6.1 through 6.6.4.

- NFPA 499 5.1.6.1 and 5.1.6.3
- NFPA 499 5.1.6.2
- NFPA 499 5.1.6.4
- NFPA 499 5.1.8.1
- NFPA 499 5.1.9
- NFPA 499 5.2.1 and 5.2.2

The items impacting electrical design and installation from NFPA 497 are incorporated into the NEC. Therefore, once the classification is known, or provided in design documentation, the applicable sections of the NEC are then applied to the electrical design—NEC Art. 500, 502, and 506.

Article 510 Hazardous (Classified) Locations—Specific

Such locations covered include commercial garages, repair and storage per Article 511, aircraft hangers in Article 513, fuel dispensing facilities in Article 514, bulk storage plants in Article 515, and so on through mobile homes (Art. 550), recreational vehicles (Art. 551), and marinas (Art. 555).

Of potential importance to an engineer who occasionally works to assist a company expand or remodel an area is Art. 590 for temporary installations.

Key Takeaways for Exam Preparation
Chapter 5 of the NEC 2023 focuses on "Special Occupancies" and is based on the National Fire Protection Association (NFPA) 70, 2023. The chapter covers various articles that provide

guidelines for electrical installations in hazardous locations, classified areas, and specific conditions.

The hazardous (classified) locations are divided into Classes I, II, and III with Divisions 1 and 2. The classification depends on the nature of the materials present, and the corresponding standards such as NFPA 497 and NFPA 499 provide classification criteria. These standards impact electrical design, and compliance with NEC Art. 500, 501, 502, 503, and 505 is essential.

Different classes and divisions have specific requirements. For Class I locations, there are additional zones based on the presence of flammable gases or vapors. Intrinsically safe systems, covered in Art. 504, ensure electrical circuits are safe from causing ignition in hazardous atmospheres.

Understanding material groups, atmospheric groups, and zone classifications is crucial. Tables and classifications provided in NFPA standards guide the design process, and engineers should focus on NEC requirements for various locations and conditions.

Hazardous Location Classification: Aerosol Products: NFPA 30B addresses hazards related to aerosol products. Compliance with NEC Arts. 500, 501, 504, and 505 is necessary when aerosols are in areas with flammable materials. Limiting sources of ignition is crucial, and electrical design should align with NEC requirements for hazardous locations.

Hazardous Location Classification: Liquids, Gases, and Vapors: NFPA 497 focuses on hazards related to liquids, gases, and vapors. Classifications depend on flash points, and the MESG or MIC ratio determines the groups. Class 1 zone requirements are applied based on material mixing with air, and engineers must follow NEC Arts. 500, 501, 502, 503, and 505 for electrical design.

Hazardous Location Classification: Combustible Dust: NFPA 499 addresses combustible dust in chemical process areas, classifying them into Groups E, F, and G. Designing electrical systems for both Division I and Division 2 locations requires adherence to NEC Arts. 500, 502, and 506. Understanding the properties of combustible dust and proper classification are essential for safe electrical installations.

Article 510 Hazardous (Classified) Locations—Specific: Specific locations like commercial garages, aircraft hangars, fuel dispensing facilities, and more are covered in this article. Engineers should be familiar with these specific requirements when dealing with installations in these areas.

Question Walkthroughs

Sample Questions

1. What is crucial to limit in areas with aerosols, according to NFPA 30B?

    a. Noise pollution

    b. Static electricity

    c. Air pollution

    d. Temperature fluctuations

2. Which NFPA standard focuses on liquids, gases, and vapors in chemical process areas?

    a. NFPA 497

    b. NFPA 499

    c. NFPA 30B

    d. NFPA 70

3. How are combustible dusts divided in NFPA 499?

    a. Groups A, B, C, D

    b. Groups IIA, IIB, IIC

    c. Groups E, F, G

    d. Groups IIIA, IIIB, IIIC

Answer and Explanation

1. **Answer: b.** Static electricity

Explanation: Limiting sources of ignition, such as static electricity, is crucial.

2. **Answer: a.** NFPA 497

Explanation: NFPA 497 focuses on liquids, gases, and vapors in chemical process areas.

3. **Answer: c.** Groups E, F, G

Explanation: Combustible dusts are divided into Groups E, F, and G.

Self-Assessment Questions

Practice Questions

1. What is the focus of Chapter 5 of the NEC 2023?

   a. Residential Installations

   b. Special Occupancies

   c. Industrial Machinery

   d. Outdoor Lighting

2. Which organization's standards is Chapter 5 based on?

   a. International Electrotechnical Commission (IEC)

   b. National Fire Protection Association (NFPA)

   c. Occupational Safety and Health Administration (OSHA)

   d. American National Standards Institute (ANSI)

3. In Article 500, what are Hazardous (Classified) Locations defined by?

   a. Presence of water

   b. Presence of explosive or ignitable mixtures

   c. High voltage requirements

   d. Structural stability

4. What does Article 501 of Chapter 5 cover?

   a. Hazardous liquid storage

   b. Class I Locations

   c. Intrinsically Safe Systems

   d. Zone 0, 1, and 2 Locations

5. What is the focus of Article 502 in Chapter 5?

   a. Flammable fibers or flyings

   b. Combustible dust

   c. Intrinsically Safe Systems

    d. Zone 0, 1, and 2 Locations

6. Which article provides guidelines for Intrinsically Safe Systems?

    a. Article 500

    b. Article 501

    c. Article 502

    d. Article 504

7. What do Zone 0, 1, and 2 Locations refer to in Article 505?

    a. Residential zones

    b. High-voltage zones

    c. Explosive gas atmosphere zones

    d. Underground zones

8. What is VRLA an abbreviation for?

    a. Very Rapid Lightning Action

    b. Volatile Resistance Limitation Apparatus

    c. Valve-Regulated Lead Acid

    d. Voltage Regulation and Load Analysis

9. Who determines the hazardous location classification according to NEC Art. 500?

    a. Electrical engineers

    b. The National Electrical Code (NEC)

    c. International Electrotechnical Commission (IEC)

    d. National Fire Protection Association (NFPA)

10. What does NFPA 30B focus on in relation to hazardous locations?

    a. Flammable aerosol products

    b. Hazardous waste disposal

    c. Electrical circuit design

    d. Structural stability

11. Which Article in Chapter 5 covers Specific Hazardous (Classified) Locations?

    a. Article 500

    b. Article 502

    c. Article 510

    d. Article 514

12. What is important for an engineer dealing with temporary installations, according to Key Takeaways?

    a. Knowledge of residential wiring

    b. Familiarity with outdoor lighting

    c. Awareness of specific hazardous locations

    d. Expertise in industrial machinery

Answers
    1. **Answer: b.** Special Occupancies

Chapter 5 of the NEC 2023 focuses on "Special Occupancies."

    2. **Answer: b.** National Fire Protection Association (NFPA)

Chapter 5 is based on the NFPA 70 standards.

    3. **Answer: b.** Presence of explosive or ignitable mixtures

Hazardous (Classified) Locations are areas where fire or explosion hazards may exist due to flammable gases, vapors, dust, or fibers.

    4. **Answer: b.** Class I Locations

Class I locations are where flammable gases or vapors are present in the air.

    5. **Answer: b.** Combustible dust

Class II locations are hazardous due to the presence of combustible dust.

    6. **Answer: d.** Article 504

Article 504 covers Intrinsically Safe Systems.

    7. **Answer: c.** Explosive gas atmosphere zones

These are locations with an explosive gas atmosphere.

8. **Answer: c.** Valve-Regulated Lead Acid

VRLA stands for Valve-Regulated Lead Acid.

9. **Answer: d.** National Fire Protection Association (NFPA)

The classification is left to the appropriate NFPA standards.

10. **Answer: a.** Flammable aerosol products

NFPA 30B focuses on the hazards of aerosol products.

11. **Answer: c.** Article 510

Article 510 covers Specific Hazardous (Classified) Locations.

12. **Answer: c.** Awareness of specific hazardous locations

Article 590 for temporary installations is of potential importance to an engineer dealing with temporary installations.

Detailed Exploration of NEC Chapter 6

Understanding NEC Chapter 6
Introduction

Chapter 6 of the NEC 2023, titled "Special Equipment," furnishes detailed directives governing the installation and utilization of special electrical equipment1. Derived from the National Fire Protection Association (NFPA) 70, 20231, this chapter encompasses various subjects, including:

- **Electric Signs and Outline Lighting (Article 600):** Encompassing the installation of conductors, equipment, and field wiring for electric signs, retrofit kits, and outline lighting, irrespective of voltage1.
- **Manufactured Wiring Systems (Article 604):** Offering guidelines for the secure use and installation of manufactured wiring systems1.
- **Office Furnishings (Article 605):** Providing directives for the safe installation and use of office furnishings1.
- **Cranes and Hoists (Article 610):** Supplying guidelines for the secure installation and use of cranes and hoists1.
- **Elevators, Dumbwaiters, Escalators, Moving Walks, Platform Lifts, & Stairway Chairlifts (Article 620):** Offering guidelines for the safe installation and use of these devices1.
- **Electric Vehicle Power Transfer System (Article 625):** Providing guidelines for the secure installation and use of electric vehicle power transfer systems1.
- **Electrified Truck Parking Spaces (Article 626):** Offering guidelines for the secure installation and use of electrified truck parking spaces1.
- **Electric Welders (Article 630):** Providing guidelines for the safe installation and use of electric welders1.
- **Audio Signal Processing, Amplification, & Reproduction Equipment (Article 640):** Offering guidelines for the secure installation and use of audio signal processing, amplification, and reproduction equipment1.
- **Information Technology Equipment (Article 645):** Providing guidelines for the secure installation and use of information technology equipment1.
- **Modular Data Centers (Article 646):** Offering guidelines for the safe installation and use of modular data centers1.
- **Sensitive Electronic Equipment (Article 647):** Providing guidelines for the secure installation and use of sensitive electronic equipment1.
- **Pipe Organs (Article 650):** Offering guidelines for the safe installation and use of pipe organs1.

- **X-Ray Equipment (Article 660):** Providing guidelines for the safe installation and use of X-ray equipment1.
- **Induction and Dielectric Heating Equipment (Article 665):** Offering guidelines for the secure installation and use of induction and dielectric heating equipment1.
- **Electrolytic Cells (Article 668):** Providing guidelines for the secure installation and use of electrolytic cells1.
- **Electroplating (Article 669):** Offering guidelines for the safe installation and use of electroplating1.
- **Industrial Machinery (Article 670):** Providing guidelines for the safe installation and use of industrial machinery1.
- **Electrically Driven or Controlled Irrigation Machines (Article 675):** Offering guidelines for the secure installation and use of electrically driven or controlled irrigation machines1.
- **Swimming Pools, Fountains, and Similar Installations (Article 680):** Providing guidelines for the safe installation and use of swimming pools, fountains, and similar installations1.
- **Natural & Artificially Made Bodies of Water (Article 682):** Offering guidelines for the secure installation and use of natural and artificially made bodies of water1.
- **Integrated Electrical Systems (Article 685):** Providing guidelines for the safe installation and use of integrated electrical systems1.
- **Solar Photovoltaic (PV) Systems (Article 690):** Offering guidelines for the secure installation and use of solar photovoltaic (PV) systems1.
- **Large-Scale Photovoltaic (PV) Electric Supply Stations (Article 691):** Providing guidelines for the safe installation and use of large-scale photovoltaic (PV) electric supply stations1.
- **Fuel Cell Systems (Article 692):** Offering guidelines for the secure installation and use of fuel cell systems1.
- **Wind Electric Systems (Article 694):** Providing guidelines for the safe installation and use of wind electric systems1.
- **Fire Pumps (Article 695):** Offering guidelines for the secure installation and use of fire pumps1.

Chapter 6 of the NEC 2023 delivers a comprehensive set of directives for the secure and effective installation and utilization of special electrical equipment, ensuring that all electrical installations adhere to the highest standards of safety and efficiency.

Core Concepts and Theories
In Chapter 6 of the 2020 NEC, "Special Equipment," and it includes the requirements for special equipment such as electric signs and outline lighting systems, electric vehicle (EV) supply equipment, solar photovoltaic (PV) systems, equipment, fire pumps and so forth. This part of the series looks at a few significant changes in articles 600 through 695.

Section 600.5 Branch Circuits

Section 600.6(A) has been revised to clarify that a sign or lighting outlet is not required at entrances for deliveries, service corridors or service hallways. Section 600.5(B) addresses markings, and it has been revised to require all disconnects supplying signs or outline lighting systems that are remotely located must be marked with the identity of the sign or lighting system it controls.

Section 600.35 Retrofit Kits

A new Section 600.35 contains requirements for retrofit kits. Retrofit kits must be listed and installed in accordance with the installation instructions. During a sign conversion, any parts found to be damaged must be replaced or repaired. A retrofitted sign must be marked to inform people that the illumination system has been replaced. Signs that are converted to tubular LED lamps powered by the existing sign sockets must include a label to alert personnel that the sign has been modified. The label must include a warning not to install fluorescent lamps.

Article 625 Electric Vehicle Power Transfer System

Article 625 has been retitled "Electric Vehicle Power Transfer System." The scope of Article 625, Section 625.1 has been revised to clarify that EVs can be connected for the purposes of charging, power export or bidirectional current flow. EVs can also now be used as a standby power source in a similar manner to the use of a standby generator or energy-storage system.

Section 625.60 AC Receptacle Outlets Used for EVPE

The term "electric vehicle power export equipment" (EVPE) is new and defined in Article 100. EVPE as defined is the electric vehicle serving as the source of electric supply. New 625.60 requires AC receptacles in electric vehicles intended to supply off-board utilization equipment to comply with 625.60. These receptacles must be listed, overcurrent protection must be provided and ground- fault circuit interrupter (GFCI) protection is required. Indication and reset capabilities for the GFCI device must be readily accessible.

680.35 Storable and Portable Immersion Pools

A new 680.35 provides requirements for storable and portable immersion pools. These pools are intended for ceremonial or ritual immersion of people. This requirement mirrors existing requirements for other pools mandating minimum distances from the pool and required GFCI protection.

680.45 Permanently Installed Immersion Pools

A new 680.45 provides requirements for permanently installed immersion pools. These pools also are intended for ceremonial or ritual immersion of people. This new requirement mirrors existing requirements for other types of pools mandating mini- mum distances, clearances and required GFCI protection.

690.12 Rapid Shutdown for PV on Buildings

The requirements in 690.12 have been revised to protect first responders. Rapid shutdown reduces the risk of electrical shock that DC/AC circuits in a PV system. To prevent PV arrays with attached inverters from hav- ing energized AC conductors within the PV array(s), the PV circuits must be specifically controlled after shutdown initiation. DC and AC circuits of PV arrays must be controlled without regard to their source of supply.

690.53 DC PV Marking

The marking requirements of 690.53 have been revised to limit the amount of information necessary, and the number of locations that must be labeled. The information on these labels must be available to qualified persons before servicing PV equipment. There are now three options for the placement of this label, including at the DC PV system disconnect, at the PV system electronic power conversion equipment or at the distribution equipment associated with the PV system.

Section 695.3 Power Source(s) for Electric Motor-Driven Fire Pumps

Section 695.3 requires fire pumps to have a reliable power source. Section 695.3(B) (1) recognizes that reliable power may not be available and permits two or more of the power sources identified in 695.3(A). The exception now permits a combination of power sources from 695.3(A), a feeder source in 695.3(C)(1) and a source in 695.3(A). Section 695.3(C)(3) has been revised to clarify that all supply-side overcurrent protective devices must be selectively coordinated.

Key Takeaways for Exam Preparation
Chapter 6 of the NEC 2023, titled "Special Equipment," is vital for understanding the safe installation and use of various electrical systems. Derived from the National Fire Protection Association (NFPA) 70, 2023, this chapter encompasses a wide array of topics, ensuring adherence to high safety standards. It covers installations like electric signs, cranes, solar PV systems, and more.

Section 600.5 Branch Circuits: The revisions in Section 600.5(A) clarify that sign or lighting outlets are not obligatory at specific entrances. Additionally, Section 600.5(B) emphasizes marking requirements for disconnects supplying signs or outline lighting systems located remotely.

Section 600.35 Retrofit Kits: A new addition, this section mandates listed retrofit kits, marking of retrofitted signs, and warnings not to install fluorescent lamps during conversions.

Article 625 Electric Vehicle Power Transfer System: This article's scope has broadened to include charging, power export, and bidirectional current flow for electric vehicles (EVs). EVs can now serve as standby power sources, similar to standby generators.

Section 625.60 AC Receptacle Outlets Used for EVPE: The introduction of "electric vehicle power export equipment" (EVPE) necessitates listed AC receptacles in EVs supplying off-board equipment to comply with specific requirements, including GFCI protection.

680.35 Storable and Portable Immersion Pools: A new requirement outlines safety measures for storable and portable immersion pools used for ceremonial or ritual immersion, mirroring existing regulations for other pools.

680.45 Permanently Installed Immersion Pools: Similar to storable and portable pools, this section provides requirements for permanently installed immersion pools intended for ceremonial or ritual use.

690.12 Rapid Shutdown for PV on Buildings: Revised to enhance first responder safety, this section requires specific control of PV circuits after shutdown initiation to prevent electrical shock risks.

690.53 DC PV Marking: Marking requirements have been simplified, offering three options for label placement and ensuring vital information accessibility to qualified personnel before servicing PV equipment.

Section 695.3 Power Source(s) for Electric Motor-Driven Fire Pumps: This section mandates a reliable power source for fire pumps, allowing flexibility by permitting a combination of power sources and requiring selective coordination of supply-side overcurrent protective devices.

Question Walkthroughs

1. Which type of pools does Section 680.35 in Chapter 6 cover?

   a. Swimming Pools

   b. Portable Immersion Pools

   c. Industrial Machinery Pools

   d. Electrically Driven Irrigation Pools

2. What is the significance of Section 625.60 AC Receptacle Outlets Used for EVPE?

   a. Marking Requirements for AC Receptacles

   b. Guidelines for Electric Vehicle Charging Stations

   c. Requirements for Electric Vehicle Power Export Equipment

   d. Safety Measures for Electric Vehicle Batteries

3. Which Section in Chapter 6 emphasizes the need for selective coordination of supply-side overcurrent protective devices for fire pumps?

   a. Section 600.5 Branch Circuits

   b. Section 625.60 AC Receptacle Outlets

   c. Section 695.3 Power Source(s) for Electric Motor-Driven Fire Pumps

   d. Section 690.53 DC PV Marking

Answer and Explanation

1. **Answer: b.** Portable Immersion Pools

Section 680.35 provides requirements for storable and portable immersion pools.

2. **Answer: c.** Requirements for Electric Vehicle Power Export Equipment

This section specifies requirements for AC receptacles in electric vehicles intended to supply off-board utilization equipment.

3. **Answer: c.** Section 695.3 Power Source(s) for Electric Motor-Driven Fire Pumps

Section 695.3(C)(3) has been revised to clarify that all supply-side overcurrent protective devices must be selectively coordinated.

Self-Assessment Questions

Practice Questions

1. What is the focus of Chapter 6 of the NEC 2023?

   a. Residential Wiring

   b. Special Electrical Equipment

   c. Outdoor Lighting

   d. Industrial Machinery

2. Which organization's standards form the basis of Chapter 6?

   a. National Geographic Society

   b. National Football League

   c. National Fire Protection Association (NFPA)

   d. National Aeronautics and Space Administration (NASA)

3. What does Article 600 of Chapter 6 cover?

   a. Office Furnishings

   b. Electric Signs and Outline Lighting

   c. Information Technology Equipment

   d. Fire Pumps

4. Which section in Chapter 6 addresses the retrofitting of signage lighting?

   a. Section 600.5 Branch Circuits

   b. Section 600.35 Retrofit Kits

   c. Section 625.60 AC Receptacle Outlets

   d. Section 680.35 Storable and Portable Immersion Pools

5. What is the revised title of Article 625 in the NEC 2023?

   a. Electric Vehicle Charging Stations

   b. Electric Vehicle Power Transfer System

   c. Electric Vehicle Safety Guidelines

d. Electric Vehicle Battery Systems

6. What is the term "EVPE" in Article 625 referred to as?

   a. Electric Vehicle Parked Engine

   b. Electric Vehicle Power Export Equipment

   c. Electric Vehicle Power Efficiency

   d. Electric Vehicle Power Enhancement

7. What does Section 690.12 aim to protect?

   a. Electrical Appliances

   b. First Responders

   c. Solar Panels

   d. Electrical Transformers

8. What is the purpose of Section 690.53 in Chapter 6?

   a. Marking DC PV systems

   b. Fire Pump Installation

   c. Rapid Shutdown for PV on Buildings

   d. Immersion Pool Requirements

9. Which Article in Chapter 6 provides guidelines for the installation of X-ray equipment?

   a. Article 625

   b. Article 660

   c. Article 670

   d. Article 680

10. What does Section 695.3 mandate for electric motor-driven fire pumps?

    a. Remote Control

    b. Reliable Power Source

    c. Solar Power Integration

    d. Wind Energy Backup

11. What is the purpose of Section 600.5(B) in Chapter 6?

   a. Installation of Electric Signs

   b. Marking Requirements for Disconnects

   c. Guidelines for Manufactured Wiring Systems

   d. Safety Measures for Cranes and Hoists

12. What does Article 646 in Chapter 6 focus on?

   a. Solar Photovoltaic Systems

   b. Modular Data Centers

   c. Sensitive Electronic Equipment

   d. Information Technology Equipment

Answers

   1. **Answer: b.** Special Electrical Equipment

Chapter 6 is specifically dedicated to guidelines for the installation and use of special electrical equipment.

   2. **Answer: c.** National Fire Protection Association (NFPA)

Chapter 6 is based on NFPA 70 standards.

   3. **Answer: b.** Electric Signs and Outline Lighting

Article 600 provides guidelines for the installation of electric signs and outline lighting.

   4. **Answer: b.** Section 600.35 Retrofit Kits

This section outlines requirements for retrofit kits in signage.

   5. **Answer: b.** Electric Vehicle Power Transfer System

Article 625 has been retitled to focus on the Electric Vehicle Power Transfer System.

   6. **Answer: b.** Electric Vehicle Power Export Equipment

EVPE stands for Electric Vehicle Power Export Equipment.

   7. **Answer: b.** First Responders

Section 690.12 aims to protect first responders by reducing the risk of electrical shock.

8. **Answer: a.** Marking DC PV systems

Section 690.53 focuses on marking requirements for DC PV systems.

9. **Answer: b.** Article 660

Explanation: Article 660 provides guidelines for X-ray equipment installation.

10. **Answer: b.** Reliable Power Source

Section 695.3 mandates a reliable power source for electric motor-driven fire pumps.

11. **Answer: b.** Marking Requirements for Disconnects

Section 600.5(B) emphasizes marking requirements for disconnects supplying signs or outline lighting systems.

12. **Answer: b.** Modular Data Centers

Article 646 provides guidelines for the installation and use of modular data centers.

Detailed Exploration of NEC Chapter 7

Understanding NEC Chapter 7

Introduction

Chapter 7 of the NEC 2023, titled "Special Conditions," furnishes detailed directives for electrical installations in specific scenarios. This chapter draws its foundation from the National Fire Protection Association (NFPA) 70, 2023.

Encompassing various subjects, the chapter includes the following:

- **Emergency Systems (Article 700):** This article outlines instructions for installing emergency systems, mandated by legal requirements and classified as emergencies by municipal, state, federal, or other codes, or any governing agency with jurisdiction. These systems automatically provide illumination or power, or both, to designated areas and equipment in case of normal electrical supply failure.
- **Legally Required Standby Systems (Article 701):** Guidelines within this article cover the installation of legally required standby systems, obligated to automatically supply power to designated areas and equipment during normal electrical supply failures.
- **Optional Standby Systems (Article 702):** This article provides guidance for the installation of optional standby systems, designed to supply power to public or private facilities where life safety is not contingent on system performance. These systems are intended to provide on-site generated power to selected loads either automatically or to manually.
- **Critical Operations Power Systems (Article 708):** Guidelines in this article pertain to the installation of critical operations power systems (COPS), designed to ensure a continuous power supply to critical operations during disruptions of the normal utility supply.
- **Interconnected Electric Power Production Sources (Article 705):** This article offers guidelines for the installation of interconnected electric power production sources, systems operating in parallel with primary electricity sources.
- **Energy Storage Systems (Article 706):** Guidelines within this article cover the installation of energy storage systems, systems storing energy from any source and providing electrical energy to a load or power grid.
- **Class 1, Class 2, & Class 3 Remote-Control, Signaling, & Power-Limited Circuits (Article 725):** This article outlines guidelines for the installation of Class 1, Class 2, & Class 3 remote-control, signaling, and power-limited circuits.
- **Fire Alarm Systems (Article 760):** This article provides directives for the installation of fire alarm systems, encompassing all circuits controlled and powered by the fire alarm system.

Chapter 7 of the NEC 2023 furnishes a comprehensive set of guidelines ensuring the safe and efficient installation of electrical systems in special conditions. Covering a diverse array of topics, these guidelines uphold the highest standards of safety and efficiency for all electrical installations.

Core Concepts and Theories
Chapter 7 of the National Electrical Code encompasses various articles, such as Emergency Systems (Article 700), Legally Required Standby Systems (Article 701), Optional Standby Systems (Article 702), Class 1, 2, and 3 Remote Control, Signaling, and Power-Limited Circuits (Article 725), Fire Alarm Circuit Wiring (Article 760), and other specialized systems.

Notably, Article 700 stands out as it addresses emergency systems without incorporating the term "standby" in its title. In contrast, Articles 701 and 702 include "standby" in their titles, placing them in a slightly lower hierarchy of systems. While the requirements in Articles 700 and 701 fall under the authority having jurisdiction, only Article 700 is exclusive to emergency systems, despite the common misconception in the electrical industry of referring to these circuits as "emergency standby circuits."

The 2020 NEC introduces several changes to Article 700 that can impact the design, application, and installation of emergency power sources and circuits, potentially simplifying the utilization of Article 700.

An important change in 700.4 involves renaming it from "Capacity" to "Capacity and Rating," with the addition of subsection (B) outlining requirements for capacity. This subsection stipulates that an emergency system must possess sufficient capacity in accordance with Article 220 or through other approved methods, which could include compliance with building codes or energy codes.

Unlike applications in certain facilities where loads may not be used simultaneously, such as hospitals, clinics, and similar healthcare facilities, an emergency system, as per 700.4, must be capable of carrying the entire emergency load at any given time. Furthermore, the system should handle any motor startups that may occur during a power outage or as part of an emergency, such as the initiation of a fire pump with a substantial startup current.

700.5(A) introduces a new statement clarifying that meter-mounted transfer switches are not permissible for emergency system use. In the early 2000s, during my tenure at Underwriters Laboratories, discussions with the manufacturer of meter-mounted transfer switches revealed their original intent for installation directly behind the utility meter and locking with the meter ring by the power company. These transfer switches were not designed for emergency power usage and were typically added by the power company.

Another addition, found in 700.5(C), specifies that transfer switches cannot be reconditioned and rebuilt, establishing an important guideline for maintaining the reliability and safety of emergency systems.

The inclusion of new subsections and titles enhances user-friendliness and facilitates access to information. A recently designated and titled section, 700.12(A), specifically addresses considerations related to power sources. This segment emphasizes that when choosing an emergency power source, factors such as occupancy and the type of service provided must be taken into account. For instance, evacuating a theater requires minimal time, while evacuating a larger, single- or multiple-story building takes longer, especially during power disruptions caused by issues inside or outside the building.

In tandem with 700.12(A), the newly labeled 700.12(B) addresses the design and location of emergency equipment, which was previously covered by two paragraphs lacking numbering or titles in the 2017 NEC. The stipulation is that emergency equipment must be designed and situated to minimize hazards leading to complete failure due to factors like flooding, fires, icing, and vandalism. The aftermath of Hurricane Katrina serves as an example of catastrophic failure, where heavy rain and flooding compromised many emergency systems in the region.

This section also mandates that emergency power sources outlined in 700.12(B) should be installed either in spaces fully protected by approved automatic fire protection systems or in spaces with a two-hour fire rating. This requirement applies to locations within assembly occupancies for more than 1,000 persons, buildings above 75 feet high with specific occupancies (assembly, education, residential, detention and correctional, business or mercantile), and educational occupancies with more than 300 occupants.

Section 700.12(H) introduces new installation requirements for DC microgrids utilized as emergency power sources. These microgrids must have a suitable rating and capacity to supply & maintain the total emergency load for a minimum of two hours during full-demand operation. Importantly, a DC microgrid system cannot serve as the sole power source for the emergency system if it also functions as the normal supply for the building or group of buildings; in such cases, a generator set may also be necessary.

Furthermore, the 2020 NEC's Section 700.16 incorporates subsection numbering and titling for enhanced usability. It is imperative for electrical designers, contractors, and electricians to familiarize themselves with Article 700 for effective application in their respective fields.

Chapter 7 of the NEC 2023 focuses on "Special Conditions," providing guidelines for safe electrical installations. Derived from NFPA 70, this chapter covers Emergency Systems, Legally Required Standby Systems, Optional Standby Systems, Critical Operations Power Systems, Interconnected Electric Power Production Sources, Energy Storage Systems, Remote-Control Circuits, and Fire Alarm Systems. It ensures compliance with high safety standards.

- **Emergency Systems (Article 700):** This addresses systems supplying power during failures. Recent changes emphasize capacity, considering the entire load and motor startups. Meter-mounted transfer switches are not allowed, and reconditioning is prohibited.
- **Legally Required Standby Systems (Article 701):** Focuses on automatic power supply during normal supply failure. Article 700 and 701 often confuse people, but only Article 700 deals with emergency systems.
- **Optional Standby Systems (Article 702):** Supplies power to non-life-safety areas during outages, emphasizing flexibility for public or private facilities.
- **Critical Operations Power Systems (Article 708):** Ensures continuous power for critical operations during utility disruptions.
- **Interconnected Electric Power Production Sources (Article 705):** Governs systems operating in parallel with primary electricity sources.
- **Energy Storage Systems (Article 706):** Guides installation of systems storing energy from any source and providing electrical energy to loads or the power grid.
- **Remote-Control, Signaling, and Power-Limited Circuits (Article 725):** Provides guidelines for different classes of circuits.
- **Fire Alarm Systems (Article 760):** Offers guidelines for installing fire alarm systems and associated circuits.

Critical Insights into Emergency Systems (Article 700):

- **Capacity and Rating (700.4):** Requires adequate capacity based on Article 220 or other approved methods. Building or energy codes may be considered.
- **Meter-Mounted Transfer Switches (700.5(A)):** Prohibits their use for emergency systems. These switches were not designed for emergency power usage.
- **Transfer Switches (700.5(C)):** Cannot be reconditioned or rebuilt, ensuring reliability.
- **Power Source Considerations (700.12(A)):** When selecting an emergency power source, factors like occupancy and service type must be considered.
- **Emergency Equipment Design and Location (700.12(B)):** Equipment must minimize hazards and be placed in areas protected against fires, flooding, and vandalism. Requirements are more specific than in the 2017 NEC.

- **DC Microgrids (700.12(H)):** New installation requirements for DC microgrids as emergency power sources, ensuring suitable rating, capacity, and a minimum of two hours of full-demand operation. Cannot serve as the sole power source if it's also the normal supply.

Question Walkthroughs

Sample Questions

1. Where must emergency sources of power described in 700.12(B) be installed, as per the 2020 NEC?

   A) Any occupied space

   B) Spaces with a one-hour fire rating

   C) Spaces fully protected by fire protection systems or with a two-hour fire rating

   D) Only in residential spaces

2. What are the new installation requirements for DC microgrids as emergency sources of power, according to 700.12(H)?

   A) Minimum one-hour full-demand operation

   B) Suitable rating and capacity for two hours of full-demand operation

   C) Exclusively serving as the sole power source

   D) Serving as the primary source for the building or group of buildings

3. What is the main takeaway regarding familiarity with Article 700 use?

   A) It is optional for electrical designers

   B) It is necessary for electricians only

   C) It is not required for exam preparation

   D) It is a must for any electrical designer, electrical contractor, or electrician

Answer and Explanation

1. **Answer: C)** Spaces fully protected by fire protection systems or with a two-hour fire rating

Emergency sources of power described in 700.12(B) must be installed in spaces fully protected by approved automatic fire protection systems or in spaces with a two-hour fire rating.

2. **Answer: B)** Suitable rating and capacity for two hours of full-demand operation

DC microgrids used as emergency sources of power must have a suitable rating and capacity to supply and maintain the total emergency load for not less than two hours of full-demand operation.

3. **Answer: D)** It is a must for any electrical designer, electrical contractor, or electrician

Becoming familiar with Article 700 use is emphasized as a must for any electrical designer, electrical contractor, or electrician.

Self-Assessment Questions

Practice Questions

1. What is the focus of Chapter 7 of the NEC 2023?

    A) General Electrical Installations

    B) Residential Wiring

    C) Special Conditions

    D) Commercial Lighting

2. Which article in Chapter 7 deals with systems that automatically supply illumination or power during failures and are legally required?

    A) Article 701

    B) Article 702

    C) Article 705

    D) Article 700

3. What type of systems does Article 702 cover?

    A) Emergency Systems

    B) Optional Standby Systems

    C) Legally Required Standby Systems

    D) Critical Operations Power Systems

4. What is the purpose of Critical Operations Power Systems (Article 708)?

    A) To supply power during failures

    B) To ensure continuity of power for critical operations

    C) To provide standby power

    D) To operate in parallel with primary sources

5. Which article provides guidelines for interconnected electric power production sources?

A) Article 706

B) Article 705

C) Article 701

D) Article 708

6. What is the purpose of Energy Storage Systems (Article 706)?

   A) To store electrical energy

   B) To generate power during outages

   C) To supply power to non-life-safety areas

   D) To provide power for critical operations

7. Which article deals with Class 1, Class 2, and Class 3 remote-control, signaling, and power-limited circuits?

   A) Article 725

   B) Article 760

   C) Article 705

   D) Article 708

8. What does Article 760 cover?

   A) Emergency Systems

   B) Fire Alarm Systems

   C) Optional Standby Systems

   D) Legally Required Standby Systems

9. In Article 700, what recent change has been made to the title of 700.4?

   A) "Load Requirements"

   B) "Capacity and Rating"

   C) "Power Source Considerations"

   D) "Emergency Circuits"

10. What is prohibited for emergency system use according to 700.5(A)?

A) Reconditioned transfer switches

B) Meter-mounted transfer switches

C) Manual transfer switches

D) Electronic transfer switches

11. What must an emergency system be able to handle, as mentioned in 700.4?

A) Motor shutdowns

B) Simultaneous load usage

C) Motor startups

D) Manual power restoration

12. What does 700.12(A) address?

A) Power Source Considerations

B) Emergency Equipment Design

C) Transfer Switch Requirements

D) Capacity and Rating

Answers

1. **Answer: C)** Special Conditions

Chapter 7 of the NEC 2023 focuses on "Special Conditions," providing guidelines for safe electrical installations.

2. **Answer: D)** Article 700

Article 700 addresses emergency systems that supply power during failures and are legally required.

3. **Answer: B)** Optional Standby Systems

Article 702 covers optional standby systems, which supply power to non-life-safety areas during outages.

4. **Answer: B)** To ensure continuity of power for critical operations

Article 708 focuses on Critical Operations Power Systems, ensuring continuous power for critical operations during utility disruptions.

5. **Answer: B)** Article 705

Article 705 provides guidelines for interconnected electric power production sources operating in parallel with primary sources.

6. **Answer: A)** To store electrical energy

Article 706 provides guidelines for the installation of energy storage systems, which store energy from any source.

7. **Answer: A)** Article 725

Article 725 provides guidelines for the installation of Class 1, Class 2, and Class 3 remote-control, signaling, and power-limited circuits.

8. **Answer: B)** Fire Alarm Systems

Article 760 provides guidelines for the installation of fire alarm systems, including all circuits controlled and powered by the fire alarm system.

9. **Answer: B)** "Capacity and Rating"

The title of 700.4 has been changed from "Capacity" to "Capacity and Rating."

10. **Answer: B)** Meter-mounted transfer switches

According to 700.5(A), meter-mounted transfer switches are not permitted for emergency system use.

11. **Answer: C)** Motor startups

An emergency system must be able to handle any motor startup, such as a fire pump.

12. **Answer: A)** Power Source Considerations

700.12(A) addresses power source considerations when selecting an emergency source of power.

Detailed Exploration of NEC Chapter 8

Understanding NEC Chapter 8

Introduction

Chapter 8 of the NEC 2023, titled "Communications Systems," furnishes detailed directives for the installation and utilization of communication systems1. Derived from the National Fire Protection Association (NFPA) 70, 20231, this chapter encompasses various subjects, including:

1. Community Antenna Television and Radio Distribution Systems (Article 820): Offering guidelines for the secure installation and operation of community antenna television and radio distribution systems1.

2. Network-Powered Broadband Communications Systems (Article 830): Providing guidelines for the secure installation and utilization of network-powered broadband communications systems1.

3. Fire Alarm Systems (Article 760): Supplying guidelines for the installation of fire alarm systems, covering all circuits controlled & powered by the fire alarm system1.

4. Optical Fiber Cables and Raceways (Article 770): Providing guidelines for the secure installation and usage of optical fiber cables and raceways1.

5. Class 1, Class 2, & Class 3 Remote-Control, Signaling, and Power-Limited Circuits (Article 725): Furnishing guidelines for the installation of Class 1, Class 2, and Class 3 remote-control, signaling, and power-limited circuits1.

In summary, Chapter 8 of the NEC 2023 delivers a comprehensive set of guidelines to ensure the secure and effective installation and operation of communication systems. Encompassing a broad spectrum of topics, it guarantees that all electrical installations adhere to the highest standards of safety and efficiency.

Core Concepts and Theories

Chapter 8 is standalone, meaning, the requirements of Chapters 1-7 do not apply unless specifically mentioned in Chapter 8. The title is communication systems also applies to cable television (CATV) and broadband communications aka the "internet" and its connections in the home or office. But it does not cover all building with communication equipment, refer to 90.2(B)(4) for those installations of circuits and equipment that are not. Many new sections were added to the Code in Chapter 8 and should be reviewed if updating any portion of such a system.

Building communication systems and equipment must adhere to the precise guidelines outlined in Chapter 8 of the NEC. Despite their usual operation at lower energy levels, inadequate grounding and bonding can lead to significant repercussions for both equipment and property, posing potential shock hazards. Article 770, along with the pertinent articles in Chapter 8 of the NEC, establishes distinct and specific grounding and bonding criteria for the installation of communication systems.

Grounding, in the simplest form, is the process of connecting an electrically conductive object to ground (the earth). Bonding is the process of connecting conductive objects together to establish continuity and conductivity. If a system or equipment is grounded, it is connected to the earth, and if objects are bonded, they are connected to electrically become one potential, or as close to the same potential as possible. These two processes work in unison to provide safety for communications systems, equipment and property, as well as operational grounding and protective grounding functions.

There are important general grounding and bonding rules for the communications systems addressed in Chapter 8 articles of the *Code*. The rules are intended to protect people and property from electrical hazards in normal operation and minimize differences of potential during abnormal events, such as line surges or lightning. Lightning is an unpredictable force, so meeting the *NEC* requirements is the minimum plan against damage from natural and unpredictable events. A lightning protection system in accordance with NFPA 780 provides another degree of protection above the minimum grounding protection required by the *NEC*.

Communications system grounding electrode conductors must be electrically common to the grounding electrode used for the electrical power system for system, equipment, and personnel safety. Article 800 (specifically Section 800.100) provides common rules specific to the grounding and bonding schemes for the communications systems covered in articles 770, 805, 810, 820, 830 and 840. These articles provide the specific minimum sizing requirements for grounding electrode conductors and bonding conductors installed for these systems. The minimum size conductors should be understood, along with specific rules that address bonding all grounding electrodes together to become one, electrically.

Grounding and bonding rules are intended to protect people and property from electrical hazards in normal operation and minimize differences of potential during abnormal events.

The reasons communications systems must be connected to the building power system grounding electrode are quite simple, yet such connections are not always made correctly. Using the same grounding electrode as the building electrical service keeps the conductive parts of communications equipment at or close to the same ground (the earth) potential in normal operation. In abnormal events, such as surges related to lightning strikes, the objective is to keep conductive parts of electrical power systems and limited-energy communications systems at the same potential as the potentials rise and fall. This minimizes the possibility of destructive flashover events within electronic equipment and between electrically conductive parts and equipment within buildings or structures. If the grounding conductors of a communications system are connected to an electrode separate from the building power service grounding electrode, a lightning event on or close to the building can cause conductive parts of equipment in the power system and the communications system to rise at different potentials, creating possible flashovers that can damage equipment or even cause a fire.

The 2020 *NEC* was revised regarding the structure and usability of the communications articles. In previous editions, these articles included a very significant number of redundant requirements repeated within each article, including grounding and bonding rules. Article 800 now includes general requirements that apply to and are common between articles 805, 820, 830 and 840. Similar grounding and bonding rules are applicable to each article and address requirements such as sizing of grounding electrode conductors, installation of bonding jumpers, installation of grounding electrode conductors and more. Each of these articles provides reference to the grounding and bonding requirements set forth in either section 770.100 or 800.100, as applicable.

As a reminder, Section 250.94(A) contains a general requirement to install an intersystem bonding termination (IBT) at the service equipment of a building or structure served. An IBT is also required at each separate building or structure supplied by one or more feeders or branch circuits. It must be installed in a way that leaves it accessible for connection and inspection. Section 250.94(B) provides an alternative method for connecting grounding and bonding conductors of communications systems by use of a copper or aluminum busbar not less than ¼ inch wide by 2 inches thick and long enough to accommodate the required connections.

800.53 Separation from Lightning Conductors

One of the recent stipulations dictates that, whenever feasible, there should be a minimum separation of 1.8 meters (6 feet) between lightning conductors and communication wires/cables, as well as CATV coaxial cables.

Part III Grounding Methods

In 800.100(A)(3), which is also new, it specifies that a bonding conductor or grounding electrode conductor cannot be smaller than 14 AWG nor does it have to be larger than 6 AWG. See *Informational Note Figure 800.100(B)(1)* for a visual of the terminology.

There are many restrictions on where certain types of cables may be used be it horizontally, vertically, under carpet, in a duct, in a raceway, air plenums, et cetera. See 800.154 and Tables 800.154(a), 800.154(b), and 500.154(c) for the many uses and restrictions.

Codes for the various cables are given in 800.179 and repeated in Table 4.

Cable Marking	Type of Use
CMP	Communications Plenum Cable
CMR	Communications Riser Cable
CMG	Communications General-Purpose Cable
CM	Communications General-Purpose Cable
CMX	Communications Limited Use
CMUC	Communications Under-Carpet

Table 4: Communication Table Types

Key Takeaways for Exam Preparation
The eighth chapter of the NEC 2023 is dedicated to the setup and utilization of communication systems. It addresses diverse subjects such as Community Antenna Television & Radio Distribution Systems, Network-Powered Broadband Communications Systems, Fire Alarm Systems, Optical Fiber Cables and Raceways, and Class 1, Class 2, and Class 3 Remote-Control, Signaling, and Power-Limited Circuits. The primary goal of this chapter is to guarantee that electrical installations adhere to rigorous safety and efficiency standards.

When diving into the core concepts, it's essential to note that Chapter 8 is standalone, meaning the rules from Chapters 1-7 don't apply unless specified. It encompasses cable television (CATV), broadband communications (internet), and their connections in homes or offices. However, it doesn't cover all buildings with communication equipment, so refer to 90.2(B)(4) for those installations not covered. Several new sections have been added to Chapter 8, necessitating a thorough review if updating any part of a system.

Grounding and bonding are critical in communication systems. Grounding connects objects to the earth, while bonding connects conductive objects to establish continuity and conductivity. Improper grounding and bonding can lead to severe consequences, so adherence to specific guidelines in Article 770 and Chapter 8 is crucial. Lightning protection systems, as outlined in NFPA 780, offer an additional layer of defense beyond NEC requirements.

Communications system grounding electrode conductors must be electrically common with the grounding electrode used for the electrical power system to ensure safety for systems, equipment, and personnel. Article 800, especially Section 800.100, outlines common rules for grounding and bonding schemes in communication systems. Understanding minimum sizing requirements for grounding electrode conductors and bonding conductors is vital.

Connecting communications systems to the building power system grounding electrode is crucial for maintaining the same ground potential in normal operation and minimizing potential differences during abnormal events like lightning strikes. Failure to connect to the same grounding electrode can result in different potentials, leading to potential flashovers and equipment damage.

The 2020 NEC introduced revisions for better structure and usability of communication articles. Article 800 now includes general requirements applicable to articles 805, 820, 830, and 840, reducing redundancy. Section 250.94(A) emphasizes the installation of an intersystem bonding termination (IBT) at the service equipment and separate buildings, providing alternatives for connecting grounding and bonding conductors of communication systems.

A notable new requirement is maintaining a separation of at least 6 feet between lightning conductors and communication wires/cables and CATV coaxial cables when practicable. In Part III Grounding Methods (800.100(A)(3)), specific guidelines regarding bonding and grounding conductor sizes are provided, ensuring they meet safety standards.

Restrictions on cable usage, whether horizontally, vertically, under carpet, in a duct, or air plenums, are outlined in 800.154 and corresponding tables. Codes for various cables are provided in 800.179 and summarized in Table 4, specifying the type of cable and its designated use.

Question Walkthroughs

Sample Questions

1. Which section provides common rules for grounding and bonding schemes in communication systems?

    A) Section 800.53

    B) Section 800.100

    C) Section 800.154

    D) Section 800.179

2. What does Article 800 encompass according to the 2020 NEC revisions?

    A) General requirements

    B) Lighting systems

    C) Heating systems

    D) Power distribution

3. What is the separation requirement between lightning conductors and communication wires/cables according to the new requirement?

    A) 2 feet

    B) 4 feet

    C) 6 feet

    D) 8 feet

Answer and Explanation

1. **Answer: B**

Section 800.100 provides common rules for grounding and bonding in communication systems.

2. **Answer: A**

Article 800 now includes general requirements applicable to various communication articles.

3. **Answer: C**

A separation of at least 6 feet is required between lightning conductors and communication wires/cables.

Self-Assessment Questions

Practice Questions

1. What is the primary focus of Chapter 8 of the NEC 2023?

   A) Lighting systems

   B) Communication systems

   C) Power distribution

   D) Heating and cooling systems

2. Which organization's standards is Chapter 8 based on?

   A) International Code Council (ICC)

   B) National Fire Protection Association (NFPA)

   C) Institute of Electrical and Electronics Engineers (IEEE)

   D) American Society of Heating, Refrigerating and Air-Conditioning Engin. (ASHRAE)

3. Which article provides guidelines for the installation of fire alarm systems?

   A) Article 830

   B) Article 770

   C) Article 725

   D) Article 760

4. What does bonding achieve in communication systems?

   A) Connecting to the earth

   B) Establishing continuity and conductivity

   C) Reducing potential differences

   D) Providing power to equipment

5. Why is it crucial for communications system grounding electrode conductors to be common with the electrical power system grounding electrode?

   A) To save energy

   B) To minimize potential differences

C) To increase conductivity

D) To reduce equipment weight

6. What is the purpose of an intersystem bonding termination (IBT) according to Section 250.94(A)?

   A) To increase power supply

   B) To connect communication systems

   C) To bond conductors

   D) To protect against lightning

7. What is the significance of maintaining the same ground potential for communication systems during normal operation?

   A) To increase efficiency

   B) To prevent flashovers

   C) To reduce cable costs

   D) To improve signal quality

8. What is the alternative method mentioned in Section 250.94(B) for connecting grounding and bonding conductors of communication systems?

   A) Using copper or aluminum busbar

   B) Using steel rods

   C) Using PVC pipes

   D) Using fiber optic cables

9. What is the minimum and maximum size specified for bonding or grounding conductors in 800.100(A)(3)?

   A) 10 AWG to 2 AWG

   B) 12 AWG to 4 AWG

   C) 14 AWG to 6 AWG

   D) 16 AWG to 8 AWG

10. Where can restrictions on cable usage be found?

A) Section 800.53

B) Section 800.100

C) Section 800.154

D) Section 800.179

11. What information is provided in Table 4?

A) Grounding methods

B) Cable types and uses

C) Lightning protection

D) General requirements

12. Which article provides guidelines for the safe installation and use of optical fiber cables and raceways?

A) Article 820

B) Article 830

C) Article 760

D) Article 770

Answers

1. **Answer: B**

Chapter 8 focuses on the installation and use of communication systems.

2. **Answer: B**

Chapter 8 is based on the standards of the National Fire Protection Association (NFPA).

3. **Answer: D**

Article 760 provides guidelines for fire alarm systems.

4. **Answer: B**

Bonding connects conductive objects to establish continuity and conductivity.

5. **Answer: B**

This ensures safety for systems, equipment, and personnel by minimizing potential differences.

6. **Answer: D**

IBT is installed to protect against lightning at the service equipment.

7. **Answer: B**

Maintaining the same ground potential prevents potential flashovers during abnormal events.

8. **Answer: A**

The alternative method involves using a copper or aluminum busbar.

9. **Answer: C**

The bonding or grounding conductor cannot be smaller than 14 AWG nor larger than 6 AWG.

10. **Answer: C**

Restrictions on cable usage can be found in Section 800.154.

11. **Answer: B**

Table 4 provides information about cable types and their designated uses.

12. **Answer: D**

Article 770 provides guidelines for optical fiber cables and raceways.

Detailed Exploration of NEC Chapter 9

Understanding NEC Chapter 9

Introduction

Chapter 9 of the National Electrical Code (NEC) for the year 2023 delves into critical aspects of electrical installations, offering a comprehensive guide to ensure the safety & efficiency of electrical systems. This chapter is essential for electricians, engineers, and professionals involved in designing, installing, and maintaining electrical systems.

- **Table 1: Percent of Cross Section of Conduit & Tubing for Conductors and Cables:** This table provides crucial information regarding the percent of cross-sectional area occupied by conductors and cables within conduits and tubing. Understanding these values is vital for proper conduit sizing, preventing issues such as overheating, and ensuring compliance with safety standards. Electricians refer to this table to determine the appropriate conduit size based on the number and size of conductors.

- **Informative Annex A: Product Safety Standards:** Annex A serves as a valuable reference for product safety standards related to electrical installations. These standards play a pivotal role in ensuring that electrical components and devices meet the required safety criteria. Electricians and manufacturers consult this annex to stay informed about the latest safety standards and to ensure the products they use or produce comply with industry regulations.

- **Informative Annex B: Application Information for Ampacity Calculation:** A crucial aspect of electrical system design is ampacity, which refers to the highest current a conductor can handle without surpassing its temperature limit. Annex B furnishes application-specific details essential for accurately determining ampacity. It covers factors like ambient temperature, conductor insulation type, and the grouping of multiple conductors. Electricians rely on this annex for precise calculations to ensure safe and efficient current-carrying capabilities.

- **Informative Annex C: Conduit, Tubing, & Cable Tray Fill Tables for Conductors & Fixture Wires of the Same Size:** Annex C offers valuable fill tables for conduits, tubing, and cable trays, specifically addressing situations where conductors and fixture wires are of the same size. Proper fill calculations are crucial to avoid overloading conduits and cable trays, which can lead to increased heat and potential safety hazards. Electricians use the information in this annex to ensure compliance with fill requirements and maintain the integrity of the electrical system.

In summary, Chapter 9 of the NEC 2023 provides essential guidance on conduit sizing, product safety standards, ampacity calculations, and fill tables. Adhering to the information in this chapter is crucial for creating electrical installations that meet safety standards, promote efficiency, and ensure the reliable operation of electrical systems.

Core Concepts and Theories

Chapter 9 is a mandatory part of the NEC. Since the metric system is worldwide except in the US, the tables for conduit list a metric trade designator and trade size. These can be found in Table 300.1(C). For example, a ¾-inch conduit metric designator 21 and trade size ¾.

One of the more commonly used tables during construction or remodeling is **Table 1 Percent of Cross Section of conduit and tubing for Conductors and Cables**. The remaining tables provide the allowable fill for the different types of conduit. Since most conduit carries two or more wires, the author recommends remembering one number from the table for immediate recall: 40%. That is, for two or more wires in the conduit the area filled cannot exceed 40%.

General guidance, not part of the NEC itself is something called the *jam ratio*. The jam ratio is defined as follows.

$$R_{jam} = \left(\frac{ID_{raceway\ (or\ conduit)}}{OD_{conductor}} \right)$$

To avoid jams in conduits or raceways, avoid values between 2.8 and 3.2.

Annexes

Recall that all annexes are not part of the NEC per se, meaning they are not requirements.

- **Annex A:** Annex A contains an extensive list of Product Safety Standard references with very specific applications.
- **Annex B:** Annex B provides guidance and examples for ampacity calculations.
- **Annex C:** Annex C contains fill tables for various configurations.
- **Annex D:** *This annex just might be the most useful in that it contains example calculations directed related to dwelling requirements.*
- **Annex E:** This annex contains construction guidance for buildings. For those requiring more information, see *NFPA 5000, Building Construction and Safety Code*.
- **Annex F:** This section, titled "Availability and Reliability for Critical Operations Power Systems," covers the development and implementation of Functional Performance Tests (FPTs) for Critical Operations Power Systems. It is particularly relevant for those overseeing standby power systems.
- For those responsible for standby power systems, this is the appropriate annex.
- **Annex G:** This annex covers *Supervisory Control and Data Acquisition (SCADA)* systems. One important suggestion given is that the COPS loads be separate from the rest of the building.

- **Annex H:** This annex is meant to be a template for local jurisdictions adopting the NEC. It should be noted that some jurisdictions will adopt the code and modify, delete, or add requirements. Those modifications, of whatever type must be understood when building in an area of a given AHJ (Authority Having Jurisdiction).
- **Annex I:** This annex contains recommended tightening torque tables from UL Standard 468A-486B.
- **Annex J:** This annex contains guidance to meet ADA standards in buildings.
- **Annex K:** This annex guides the use of Medical Electrical Equipment in Dwellings & Residential Board-and-Care Occupancies.

Key Takeaways for Exam Preparation

Chapter 9 of the NEC 2023 delves into various critical aspects, including Table 1, which outlines the percentage of cross-sections for conduits and tubing for conductors and cables. Additionally, it includes informative annexes, such as Annex A, covering Product Safety Standards, Annex B providing application guidance for ampacity calculations, and Annex C presenting fill tables for conduits, tubing, and cable trays. Understanding the core concepts and theories in this chapter is essential, especially considering its mandatory nature within the NEC.

Chapter 9, a vital section of the NEC, employs the metric system for conduit specifications, a global standard with exceptions in the US. Table 300.1(C) is a key reference for metric trade designators and trade sizes, making it crucial for practical applications. The significance of Table 1, determining the percentage of cross-sections for conduits and tubing, cannot be overstated. Notably, a fundamental recommendation is to memorize the 40% limit for conduit fill when accommodating two or more wires. Understanding the jam ratio, avoiding values between 2.8 and 3.2 to prevent conduit jams, is a valuable practical tip not explicitly part of the NEC itself.

Annexes in Chapter 9 are informative supplements rather than mandatory requirements. Annex A provides an extensive list of Product Safety Standard references with specific applications. Annex B offers essential guidance and examples for ampacity calculations, while Annex C contains fill tables for various configurations. Annex D proves highly practical, offering example calculations directly related to dwelling requirements. Annex E provides construction guidance for buildings, directing those seeking more information to NFPA 5000, the Building Construction and Safety Code. Annex F, aptly titled "Availability and Reliability for Critical Operations Power Systems," is a must-read for those managing standby power systems. Annex G focuses on Supervisory Control and Data Acquisition (SCADA) systems, emphasizing the separation of COPS loads from the rest of the building. Annex H serves as a template for local jurisdictions adopting the NEC, highlighting the need to understand modifications made by authorities. Annex I contains recommended tightening torque tables, and Annex J offers guidance to meet ADA standards in buildings. Lastly, Annex K provides direction for the use of Medical Electrical Equipment in Dwellings and Residential Board-and-Care Occupancies.

Thorough preparation for the exam involves a comprehensive understanding of Chapter 9 and its tables, with a focus on practical applications and recommendations. Annexes offer valuable supplementary information, ensuring a well-rounded grasp of the NEC 2023. Exam takers should prioritize key concepts, such as conduit fill limits, metric system applications, and the significance of informative annexes, to navigate the exam successfully.

Question Walkthroughs

Sample Questions

1. What is the recommended percentage limit for conduit fill when accommodating two or more wires?

   A) 20%

   B) 40%

   C) 60%

   D) 80%

2. What does the jam ratio refer to in electrical installations?

   A) Metric system conversion

   B) Conduit fill calculations

   C) Avoiding conduit jams

   D) Cable tray configurations

3. In which table can you find the metric trade designator and trade size for conduits?

   A) Table 1

   B) Table 300.1(C)

   C) Annex A

   D) Annex B

Answer and Explanation

1. Answer: B

Explanation: The recommended limit for conduit fill with two or more wires is 40%.

2. Answer: C

Explanation: The jam ratio helps avoid jams in conduits or raceways, particularly values between 2.8 and 3.2.

3. Answer: B

Explanation: The metric trade designator and trade size for conduits can be found in Table 300.1(C).

Self-Assessment Questions

Practice Questions

1. What is the primary purpose of Chapter 9 in the NEC 2023?

   A) Provide construction guidelines

   B) Outline product safety standards

   C) Present conduit fill tables

   D) Define metric trade designators

2. Which annex provides an extensive list of Product Safety Standard references?

   A) Annex A

   B) Annex B

   C) Annex C

   D) Annex D

3. Which annex offers guidance and examples for ampacity calculations?

   A) Annex A

   B) Annex B

   C) Annex C

   D) Annex D

4. Which annex contains fill tables for various conduit configurations?

   A) Annex A

   B) Annex B

   C) Annex C

D) Annex D

5. Which annex contains example calculations directly related to dwelling requirements?

A) Annex D

B) Annex E

C) Annex F

D) Annex G

6. For what purpose is Annex F primarily useful?

A) Construction guidance

B) Standby power systems

C) SCADA systems

D) Tightening torque tables

7. What does Annex G focus on?

A) Building construction

B) ADA standards

C) SCADA systems

D) Medical Electrical Equipment

8. What does Annex H serve as for local jurisdictions adopting the NEC?

A) Construction guide

B) Template

C) Safety standard

D) Ampacity calculator

9. Which annex contains recommended tightening torque tables?

A) Annex I

B) Annex J

C) Annex K

D) Annex H

10. Which annex provides guidance to meet ADA standards in buildings?

    A) Annex A

    B) Annex B

    C) Annex J

    D) Annex K

11. What does Annex K provide guidance for?

    A) Construction guidance

    B) Ampacity calculations

    C) Medical Electrical Equipment

    D) SCADA systems

12. What is crucial for exam takers to prioritize for successful navigation of the Chapter 9 exam?

    A) Memorizing all tables

    B) Understanding core concepts and theories

    C) Focusing on metric system conversions

    D) Ignoring informative annexes

Answers

1. **Answer: C**

Chapter 9 focuses on tables, including conduit fill tables, in the NEC 2023.

2. **Answer: A**

Annex A contains an extensive list of Product Safety Standard references.

3. **Answer: B**

Annex B provides guidance and examples for ampacity calculations.

4. **Answer: C**

Annex C contains fill tables for various conduit configurations.

5. **Answer: A**

Annex D contains example calculations related to dwelling requirements.

6. **Answer: B**

Annex F is useful for those managing standby power systems.

7. **Answer: C**

Annex G covers Supervisory Control and Data Acquisition (SCADA) systems.

8. **Answer: B**

Annex H serves as a template for local jurisdictions adopting the NEC.

9. **Answer: A**

Annex I contains recommended tightening torque tables.

10. **Answer: C**

Annex J provides guidance to meet ADA standards in buildings.

11. **Answer: C**

Annex K provides guidance for the use of Medical Electrical Equipment.

12. **Answer: B**

Understanding core concepts and theories, including conduit fill limits and the metric system, is crucial for exam success.

PART III: REVIEW, RECALIBRATE, AND RISE

Synthesizing Knowledge and Strategizing Revision

Bringing It All Together

The National Electrical Code (NEC) serves as the cornerstone for electrical safety standards in the United States. Comprising multiple chapters, each addressing specific aspects of electrical installations, the NEC provides a comprehensive framework for safeguarding lives and property. This holistic view explores the key elements of each chapter, illustrating how they interconnect to form a cohesive electrical safety strategy.

Chapter 1: General Requirements

The NEC begins with Chapter 1, establishing fundamental principles. This chapter covers definitions, responsibilities, and general requirements applicable to all installations. It sets the stage for subsequent chapters by emphasizing the importance of compliance and safety awareness.

Chapter 2: Wiring and Protection

Chapter 2 delves into the specifics of wiring and protection. It addresses conductor sizing, overcurrent protection, and grounding. These elements form the backbone of a safe electrical system, ensuring that conductors are appropriately sized to handle loads, overcurrents are mitigated, and grounding is established to prevent electrical shock hazards.

Chapter 3: Wiring Methods and Materials

Building on Chapter 2, Chapter 3 provides guidelines for wiring methods and materials. It explores the various options available for routing and securing conductors, emphasizing the importance of choosing methods suitable for specific environments. This chapter promotes flexibility and adaptability while maintaining safety standards.

Chapter 4: Equipment for General Use

Chapter 4 focuses on the equipment used in electrical installations. From switches and receptacles to appliances and luminaires, this chapter addresses the design, installation, and protection of general-use electrical equipment. Compliance ensures that these components function safely within the overall electrical system.

Chapter 5: Special Occupancies

Certain occupancies pose unique electrical challenges. Chapter 5 covers special occupancies such as hazardous locations, health care facilities, and agricultural buildings. It tailors electrical requirements to accommodate the specific needs and potential risks associated with these environments, ensuring safety in diverse settings.

Chapter 6: Special Equipment

Chapter 6 addresses special equipment, including transformers, capacitors, and generators. It provides guidelines for the installation, operation, and protection of these devices, ensuring their integration into electrical systems without compromising safety or performance.

Chapter 7: Special Conditions

Chapter 7 deals with special conditions, encompassing requirements for emergency systems, legally required standby systems, and optional standby systems. It emphasizes the critical role these systems play in maintaining functionality during power outages or emergencies, reinforcing the reliability of electrical installations.

Chapter 8: Communication Systems

As technology evolves, so do the communication systems covered in Chapter 8. This chapter addresses the installation of communication circuits and equipment, recognizing the increasing importance of data transmission in modern society. It ensures that these systems integrate seamlessly with electrical installations while adhering to safety standards.

Chapter 9: Tables

Chapter 9 consolidates various tables essential for proper electrical design and installation. It includes tables for conductor ampacity, motor and transformer protection, and conduit fill, among others. Designers and installers reference these tables to make informed decisions and ensure compliance with NEC standards.

A holistic understanding of the NEC involves recognizing the interconnectedness of its chapters. From general requirements to specialized conditions and equipment, each chapter contributes to the overarching goal of electrical safety. Adherence to NEC standards is paramount in designing, installing, and maintaining electrical systems that safeguard lives and property. As technology advances and new challenges emerge, the NEC continues to evolve, providing a dynamic framework for electrical safety in the ever-changing landscape of the built environment.

Identifying Strengths and Pinpointing Weaknesses

Embarking on a journey as a Journeyman Electrician necessitates not only technical proficiency but also a reflective self-assessment to identify strengths and weaknesses. This guided self-assessment aims to pinpoint key areas of competence while acknowledging opportunities for growth.

Strengths:

- **Technical Competence:** A cornerstone of strength lies in technical expertise. A Journeyman Electrician's in-depth knowledge of electrical systems, codes, and regulations enables the safe and efficient installation, maintenance, and repair of electrical components. This proficiency is fundamental to ensuring the integrity and reliability of electrical systems.

- **Problem Solving:** A Journeyman Electrician excels in problem-solving. Whether troubleshooting a malfunctioning circuit or adapting to unexpected challenges on a job site, the ability to analyze, diagnose, and resolve issues efficiently is a valuable strength. This skill contributes not only to individual success but also to the overall productivity of the team.

- **Adherence to Safety Protocols:** A commitment to safety is a hallmark strength. Journeyman Electricians prioritize safety protocols, recognizing the inherent risks associated with electrical work. The strict adherence to safety standards not only protects oneself but also promotes a secure working environment for colleagues and clients.

- **Attention to Detail:** The devil often resides in the details, and a Journeyman Electrician's acute attention to detail is a significant strength. From precise measurements to meticulous wiring, this attention ensures the accuracy and quality of electrical installations, reducing the likelihood of errors.

- **Communication Skills:** Effective communication is a strength that extends beyond technical competence. Journeyman Electricians excel in conveying complex technical information to team members, clients, and other stakeholders. Clear communication fosters collaboration, minimizes misunderstandings, and contributes to successful project outcomes.

Weaknesses:

- **Adaptability to Emerging Technologies:** In the rapidly evolving field of electrical work, a weakness may lie in adapting to emerging technologies. As renewable energy sources, smart systems, and advanced automation become more prevalent, staying abreast of these changes is crucial. A commitment to continuous learning is essential to overcome this weakness.

- **Time Management:** Time management can be a challenge in the dynamic and fast-paced environment of electrical work. Balancing multiple tasks, adhering to project timelines, and efficiently allocating resources require honed time management skills. Developing strategies to enhance organizational skills and optimize time utilization is an opportunity for improvement.

- **Leadership Skills:** As a Journeyman Electrician advances in their career, leadership skills become increasingly important. This includes the ability to lead a team, delegate tasks, and make informed decisions. Recognizing and actively developing leadership skills can open avenues for career progression.

- **Networking and Professional Development:** Building a professional network and actively engaging in continuous professional development can be areas for improvement. Networking facilitates knowledge exchange and opens doors to new opportunities, while ongoing learning ensures relevance in a field that continually evolves.

- **Customer Relations:** Enhancing customer relations is an opportunity for growth. Journeyman Electricians often interact with clients, and cultivating strong interpersonal skills can contribute to client satisfaction and repeat business. Improving communication with clients, understanding their needs, and providing clear explanations of work performed can enhance overall customer relations.

Actionable Revision Strategies

Targeted Review Techniques

Targeted review techniques are essential for effective learning and improvement, especially when preparing for exams or seeking to reinforce knowledge. Here's a structured approach based on the points you've outlined:

1. **Answer Questions Chapter by Chapter**

Ensure a comprehensive understanding of each chapter.

- Create chapter-wise question sets or use existing practice exams.
- Answer questions without referencing materials initially.
- Review correct and incorrect answers, taking note of areas requiring reinforcement.
- Focus on the reasoning behind each answer to deepen comprehension.

2. **Focus on Wrong Answers**

Identify and address specific areas of weakness.

- Compile a list of questions answered incorrectly.
- Categorize these questions based on the corresponding chapters or topics.
- Analyze the patterns of errors (e.g., consistently misunderstanding a specific concept).
- Revisit the relevant sections of the material, seeking additional resources if necessary.
- Reattempt the incorrect questions periodically to track progress.

3. **Find Patterns to Understand Knowledge Gaps**

Uncover recurring themes or concepts causing difficulties.

- Group incorrect answers by topic or concept.
- Look for patterns in the types of questions consistently answered incorrectly.
- Identify commonalities, such as misunderstanding specific code requirements or principles.
- Create targeted study sessions focusing on these recurring themes.
- Seek clarification from instructors, online resources, or textbooks to address knowledge gaps.

4. **Going Back and Filling Knowledge Gaps**

Systematically reinforce weak areas and enhance overall understanding.

- Utilize textbooks, online resources, or notes to revisit chapters with identified knowledge gaps.
- Create concise summaries or flashcards for key concepts.
- Engage in active learning methods, such as teaching the material to someone else or solving related problems.
- Form study groups to discuss challenging topics and learn from peers.
- Regularly reassess knowledge by revisiting previously incorrect questions.

Streamlining Last-Minute Prep
Do's: (recommended practices)

- **Prioritize Key Topics:** Identify high-priority topics that are likely to appear on the exam. Focus on core concepts and frequently tested areas.
- **Practice with Mock Exams:** Take practice exams under timed conditions to simulate the real test environment. This helps improve time management and builds confidence.
- **Review Summary Notes:** Use concise summary notes or flashcards for quick reviews of essential formulas, concepts, and key points. These aids can be handy for last-minute reinforcement.
- **Clarify Doubts:** If any concepts remain unclear, seek clarification from instructors, online forums, or study groups. Addressing doubts ensures a solid understanding of the material.
- **Use Mnemonics and Memory Aids:** Employ mnemonic devices and memory aids to remember complex information. This can be particularly helpful for recalling lists, codes, or sequences.
- **Stay Hydrated and Get Adequate Rest:** Maintain good physical health. Drink water and ensure you get sufficient rest before the exam. A well-rested mind performs better under stress.
- **Create a Checklist:** Develop a checklist of materials needed for the exam day, such as identification, admission ticket, and required writing instruments.
- **Positive Visualization:** Engage in positive visualization. Envision yourself confidently answering questions and successfully completing the exam. Positive thinking can help manage stress.

Don'ts: (to avoid)

- **Cram Excessively:** Attempt to cram vast amounts of information at the last minute. Focus on reinforcing what you already know rather than trying to learn entirely new topics.
- **Overload on Caffeine:** Rely on excessive caffeine intake. While a moderate amount can help with alertness, too much can lead to jitteriness and increased anxiety.

- **Neglect Breaks:** Forget to take short breaks during study sessions. Breaks help maintain focus and prevent burnout.
- **Dwell on Unanswered Questions:** Spend too much time on a single question during practice exams. If unsure, mark it for review and move on. Dwelling on one question can eat up valuable time.
- **Procrastinate on Logistics:** Wait until the last minute to gather required materials for the exam day. Ensure you have everything well in advance to avoid unnecessary stress.
- **Stay Up All Night:** Sacrifice a night's sleep for additional study time. Sleep is crucial for cognitive function, and a tired mind is less likely to perform well during exams.
- **Engage in Negative Self-Talk:** Allow negative thoughts to dominate your mindset. Replace self-doubt with positive affirmations. Confidence is key during exams.
- **Underestimate Exam Conditions:** Ignore the importance of understanding the exam format, rules, and time constraints. Familiarize yourself with these aspects beforehand to minimize surprises on the actual day.

Tactical Problem-Solving

Encountering tricky questions during exam preparation is common, especially when studying materials like the NEC (National Electrical Code). Tackling these questions strategically requires a combination of knowledge, critical thinking, and effective problem-solving skills. Here's a tactical approach to dealing with tricky questions using your study materials:

1. **Understand the Question:** Take your time to read the question thoroughly. Identify keywords and any nuances that may impact your understanding of what is being asked.

2. **Refer to NEC2023 and Study Book:**

   NEC2023 Book:
   - **Use Index and Table of Contents:** If the question pertains to a specific code or regulation, refer to the index or table of contents in the NEC2023 book to locate the relevant section.
   - **Cross-References:** Pay attention to cross-references within the code. They may guide you to related sections that provide additional context.

   Study Book:
   - **Review Related Content:** Consult your study book for additional explanations, examples, or illustrations related to the topic in question.
   - **Check Footnotes and Annotations:** Some study materials include additional information in footnotes or annotations that can clarify concepts.

3. **Break Down the Question:** Break the question into its components. Identify what is being asked, the key variables, and any conditions or constraints. Put the question in your own words. This can sometimes make it easier to understand and solve.

4. **Eliminate Distractions:** Focus on the information that directly relates to the question. Tricky questions may include distractors, so be mindful of extraneous details.

5. **Use Logical Reasoning:** Even if you don't have the answer at your fingertips, apply logical reasoning based on your understanding of electrical principles and codes. If the question is multiple-choice, eliminate options that are clearly incorrect. This can increase your chances of selecting the correct answer.

6. **Seek Clarification:** If you encounter a question that is particularly challenging or unclear, flag it for review and come back to it later. If you have access to instructors, study groups, or forums, seek clarification on tricky concepts or questions.

7. **Practice Active Recall:** Engage in active recall by remembering relevant information related to the question. This helps reinforce your understanding of the material. Relate the question to real-world scenarios or practical applications based on your knowledge.

8. **Simulate Exam Conditions:** Simulate exam conditions during your study sessions. This includes time constraints, which can be a factor in dealing with tricky questions during the actual exam.

9. **Review and Learn:** After completing a practice exam or encountering a tricky question, review the correct answer and the explanation. Understand the rationale behind the answer to enhance your knowledge.

The Final Countdown

A Week Before the Exam

As you approach the final week before your exam, it's crucial to focus on consolidation, confidence-building, and ensuring that you're physically and mentally prepared. Here's a checklist and key considerations for the week leading up to the exam:

Key Considerations:

1. **Review High-Priority Topics:** Focus on the key topics that are frequently tested or that you find challenging. Revisit notes, textbooks, and any flagged questions from your practice exams.

2. **Simulate Exam Conditions:** Take a full-length practice exam under timed conditions. This helps you assess your time management skills and builds confidence in handling the exam format.

3. **Clarify Last-Minute Doubts:** Seek clarification on any remaining doubts or confusing concepts. Use online forums, study groups, or consult instructors to ensure a clear understanding.

4. **Mindful Relaxation:** Incorporate relaxation techniques such as deep breathing or meditation to manage stress. A calm mind enhances focus and retention.

5. **Review NEC2023:** Focus on key sections and code requirements in the NEC2023. Pay attention to any updates or changes from previous editions.

6. **Healthy Sleep Routine:** Ensure you get adequate sleep each night. A well-rested mind improves cognitive function and memory recall.

7. **Stay Hydrated and Eat Well:** Maintain good hydration and nutrition. Avoid excessive caffeine and sugar, opting for balanced meals to sustain energy levels.

8. **Plan Exam Day Logistics:** Confirm the exam location and gather all necessary materials. Ensure you have your identification, admission ticket, writing instruments, and any other required items.

9. **Positive Visualization:** Visualize yourself successfully completing the exam. Positive visualization can boost confidence and help manage anxiety.

Checklist:

1. **Exam Logistics:**

   - ☐ Verify the exam date, time, and location.
   - ☐ Prepare and organize all required materials (id card, admission ticket, pens, etc.).
   - ☐ Plan your route to the exam center and account for potential traffic or delays.

2. **Review Materials:**

   - ☐ Go through summary notes or flashcards for quick reviews.
   - ☐ Revisit challenging topics and areas identified during practice exams.
   - ☐ Skim through key sections of the NEC2023.

3. **Practice Exams:**

   - ☐ Take at least one full-length practice exam under timed conditions.
   - ☐ Review ALL answers and focus on understanding the rationale behind each.

4. **Self-Care:**

   - ☐ Ensure you are getting sufficient sleep throughout the week.
   - ☐ Maintain a healthy diet and stay hydrated.
   - ☐ Incorporate stress-relief activities such as exercise or relaxation techniques.

5. **Last-Minute Clarifications:**

   - ☐ Seek clarification on any remaining questions or concepts causing confusion.
   - ☐ Participate in study groups or online forums for collaborative learning.

6. **Final Confidence Check:**

   - ☐ Reflect on your preparation and acknowledge the effort you've put in.
   - ☐ Remind yourself of your strengths and areas of improvement.

7. **Exam Strategy:**

   - ☐ Outline a strategy for approaching the exam, including time allocation for each section.
   - ☐ Plan how you will handle challenging questions—whether to skip and return, or make an educated guess.

8. **Relaxation Techniques:**

   - ☐ Practice relaxation techniques to manage stress and promote a calm mindset.

Exam Day Insights

Approaching exam day with a clear mindset and effective strategies can greatly contribute to your success. Here are insights on what to expect, how to navigate the exam day, and tips on staying calm:

What to Expect:

1. **Check-In Process:** Be prepared for a check-in process that may include verification of your identification, admission ticket, and any required materials.

2. **Exam Environment:** Expect a controlled and quiet environment. Follow the rules and guidelines provided by the exam center.

3. **Exam Format:** Understand the format of the exam, including the number of sections, question types, and time constraints.

4. **Time Management:** Time management is crucial. Be aware of the time allocated for each section and pace yourself accordingly.

5. **Challenging Questions:** Anticipate encountering challenging questions. If you get stuck on a particular question, consider marking it for review and moving on to prevent time pressure.

6. **Adaptability:** Be adaptable. If a question seems unfamiliar, apply your problem-solving skills and logical reasoning.

How to Navigate:

1. **Read Instructions Carefully:** Start by carefully reading all instructions before beginning each section. Ensure you understand the rules and requirements.

2. **Manage Time Effectively:** Allocate time wisely. Don't spend too much time on a single question. If you're unsure, mark it for review and move on.

3. **Answer Every Question:** In most exams, there is no penalty for guessing. If you're unsure about an answer, make an educated guess. It's better to have a chance at getting it right than leaving it blank.

4. **Review Marked Questions:** If time permits, go back to review any questions you marked. Double-check your answers and make any necessary corrections.

5. **Stay Focused:** Maintain focus on the current question. Avoid thinking about previous questions or worrying about upcoming ones.

How to Stay Calm:

1. **Mindful Breathing:** Practice mindful breathing during the exam. Deep breaths can help calm nerves and improve focus.

2. **Positive Self-Talk:** Replace negative thoughts with positive affirmations. Remind yourself of your preparation and capability.

3. **Focus on the Present:** Concentrate on the question at hand. Worrying about the overall exam or future questions can lead to unnecessary stress.

4. **Take Short Breaks:** If permitted, take short breaks between sections to stretch and clear your mind. Use this time to relax and refocus.

5. **Visualize Success:** Visualize yourself successfully answering questions and completing the exam. Positive visualization can enhance confidence.

6. **Trust Your Preparation:** Trust the preparation you've done. Confidence in your knowledge and abilities can significantly reduce stress.

7. **Stay Hydrated:** Ensure you stay hydrated by taking small sips of water during breaks. Dehydration can affect concentration.

8. **Accept Imperfection:** Understand that you may not know the answer to every question, and that's okay. Focus on doing your best overall.

Post-Exam Reflection

After completing the exam, it's essential to engage in a thoughtful post-exam reflection to process your experience and plan your next steps. Here are key considerations to guide your reflection:

1. **Immediate Thoughts:** Reflect on your immediate thoughts and feelings after completing the exam. Note any areas where you felt confident and areas that presented challenges.

2. **Positive Aspects:** Identify and celebrate the positive aspects of your performance. Acknowledge questions you answered confidently and sections where your preparation paid off.

3. **Challenges and Mistakes:** Recognize the challenges you faced and any mistakes made. Avoid dwelling on specific questions but focus on learning from these experiences for future improvement.

4. **Time Management:** Evaluate your time management during the exam. Consider whether you allocated time effectively and if there were sections where you could have utilized your time more efficiently.

5. **Unfamiliar Concepts:** Reflect on any unexpected or unfamiliar concepts that appeared in the exam. This insight can guide your future study efforts and highlight areas for additional focus.

6. **Question Types:** Consider the types of questions you found difficult. This reflection can help you understand the specific skills or knowledge areas that may need strengthening.

7. **Exam Environment:** Assess how well you adapted to the exam environment. Reflect on how factors such as noise, lighting, or the physical setting may have influenced your performance.

8. **Post-Exam Emotions:** Acknowledge any post-exam emotions, whether relief, anxiety or a mix of feelings. Understanding your emotional response can provide insight into your personal exam-taking experience.

Next Steps:

1. **Wait for Results:** Be patient while waiting for the exam results. Avoid unnecessary stress by focusing on other aspects of your life and maintaining a balanced perspective.

2. **Identify Learning Opportunities:** Use your post-exam reflection to identify learning opportunities. Consider creating a list of topics or skills that require additional attention for future improvement.

3. **Plan for Future Preparation:** Based on your reflection, develop a plan for future preparation. This could include adjusting study methods, seeking additional resources, or focusing on specific content areas.

4. **Celebrate Achievements:** Celebrate the successes of your exam preparation. Recognize the effort you invested and the knowledge you gained throughout the process.

5. **Maintain a Growth Mindset:** Cultivate a growth mindset by viewing challenges as opportunities for learning and improvement. Embrace the journey of continuous learning in your field.

6. **Stay Positive:** Regardless of the outcome, maintain a positive outlook. Understand that exams are part of the learning process, and each experience contributes to your growth and development.

Continuous Learning and Career Navigation

Embarking on the Journey Post-Exam

Embarking on the journey post-exam opens up a spectrum of opportunities and pathways for aspiring electricians, laying the foundation for advancement in the field. One promising avenue is the pursuit of becoming a Master Electrician, representing the pinnacle of expertise and leadership in electrical work. Achieving the status of a Master Electrician involves acquiring additional experience, often measured in years, and meeting specific licensing requirements. This journey includes honing skills in complex installations, managing electrical projects, and demonstrating a comprehensive understanding of electrical codes and regulations. Aspiring Master Electricians typically engage in continuous learning, staying abreast of technological advancements, and taking on supervisory roles to develop their leadership capabilities. Additionally, pursuing advanced certifications, such as the Master Electrician license, not only elevates one's professional standing but also opens doors to more diverse and challenging opportunities within the electrical industry. The post-exam period serves as a stepping stone toward this ambitious goal, motivating individuals to refine their craft, contribute to the safety and efficiency of electrical systems, and ultimately shape the future of their careers as accomplished Master Electricians.

Staying Updated

Continuous learning is imperative in the dynamic field of electrical work, and staying updated with industry advancements is crucial for professional growth. Engaging in various learning avenues ensures that electricians remain well-informed about the latest technologies, codes, and best practices. Here are effective ways to stay updated:

Online Communities:

- **Participate in Forums:** Join online forums and communities dedicated to electrical work. Platforms like Reddit or professional forums provide spaces for discussions, problem-solving, and knowledge sharing.
- **Ask Questions:** Actively participate by asking questions and seeking advice. The collective wisdom of the community can provide valuable insights and solutions to real-world challenges.

YouTube and Online Courses:

- **Subscribe to Channels:** Subscribe to reputable YouTube channels that focus on electrical work. Many experienced electricians and organizations share tutorials, demonstrations, and insights into the latest technologies.
- **Online Courses:** Enroll in online courses from reputable platforms. These courses often cover advanced topics, new technologies, and changes in regulations.

Follow Industry Publications:

- **Subscribe to Journals:** Subscribe to electrical industry journals and publications. These sources provide in-depth articles, case studies, and updates on emerging technologies and trends.
- **Read Newsletters:** Sign up for newsletters from industry organizations. These newsletters often contain curated content, including the latest developments and upcoming events in the field.

Attend Workshops and Seminars:

- **Local and Virtual Events:** Attend workshops, seminars, and conferences either in person or virtually. These events offer opportunities to learn from experts, engage with industry leaders, and explore new products and technologies.

Networking:

- **Join Professional Associations:** Become a member of professional associations related to electrical work. These organizations often host events and provide networking opportunities, enabling you to connect with peers and mentors.
- **LinkedIn:** Use LinkedIn to connect with professionals in the electrical industry. Follow relevant companies and join groups to stay updated on industry news and discussions.

Read Industry Codes and Standards:

- **Regularly Review NEC:** Stay updated on changes to the National Electrical Code (NEC) and other relevant standards. Familiarize yourself with the latest editions and amendments to ensure compliance in your work.
- **Local Codes:** Be aware of local electrical codes and regulations, as these may differ from the national standards. Regularly check for updates and revisions.

Utilize Technology:

- **Apps and Software:** Explore apps and software tools designed for electricians. These tools can assist in calculations, provide quick references, and streamline various aspects of electrical work.
- **Webinars and Podcasts:** Attend webinars and listen to podcasts hosted by industry experts. These platforms often cover timely topics and offer insights into current trends and challenges.

Navigating Career Progression

A Journey Electrician's career evolution involves a strategic approach to advancing and diversifying their skill set, paving the way for increased responsibilities and opportunities. After gaining proficiency in the fundamentals, one pathway to career progression is pursuing specialized certifications or endorsements in areas such as renewable energy, automation, or energy management. This not only broadens the skill set but also aligns the electrician with emerging industry trends. Additionally, taking on leadership roles within projects or teams enhances supervisory and managerial skills. Engaging in continuing education programs, workshops, and industry conferences ensures that the electrician stays current with evolving technologies and regulations. As experience grows, some may opt to pursue licensure as a Master Electrician, enabling them to oversee and manage more complex projects independently. Diversifying expertise in areas like smart home technology, electric vehicle charging, or industrial control systems can position a Journey Electrician as a versatile professional in high-demand sectors. Moreover, venturing into entrepreneurship by establishing an electrical contracting business offers a unique avenue for career autonomy and business ownership. The key to successful career progression lies in a commitment to lifelong learning, adaptability to industry changes, and a proactive approach to skill development, ultimately positioning the electrician for a dynamic and fulfilling professional journey.

Appendices

A. Sample Study Schedules
Below is a sample study schedule template in table format. Adjust the times, subjects, and durations based on your preferences, availability, and specific study needs.

Time	Monday	Tuesday	Wednesday	Thursday	Friday	Saturday
8:00 AM	Wake up	Wake up	Wake up	Wake up	Wake up	Wake up
8:30 AM	Breakfast	Breakfast	Breakfast	Breakfast	Breakfast	Breakfast
9:00 AM	Chapter 1	Chapter 1	Chapter 4	Chapter 4	Chapter 7	Chapter 7
11:00 AM	Break	Break	Break	Break	Break	Break
11:15 AM	Chapter 2	Chapter 2	Chapter 5	Chapter 5	Chapter 8	Chapter 8
1:00 PM	Lunch	Lunch	Lunch	Lunch	Lunch	Lunch
2:00 PM	Chapter 3	Chapter 3	Chapter 6	Chapter 6	Chapter 9	Chapter 9
4:00 PM	Break/Exercise	Break/Exercise	Break/Exercise	Break/Exercise	Break/Exercise	Break/Exercise
4:30 PM	Practice Questions	Practice Questions	Practice Questions	Practice Questions	Practice Questions	Practice Questions
6:00 PM	Dinner	Dinner	Dinner	Dinner	Dinner	Dinner
7:00 PM	Review/Flashcards	Review/Flashcards	Review/Flashcards	Review/Flashcards	Review/Flashcards	Review/Flashcards
8:00 PM	Relax/Free Time	Relax/Free Time	Relax/Free Time	Relax/Free Time	Relax/Free Time	Relax/Free Time
9:00 PM	Wind Down	Wind Down	Wind Down	Wind Down	Wind Down	Wind Down
10:00 PM	Bedtime	Bedtime	Bedtime	Bedtime	Bedtime	Bedtime

B. Recommended Books, Resources, and Procurement Sites
For Journeyman Electricians, there are several recommended books, resources, and procurement sites that can aid in skill development, exam preparation, and staying current with industry standards. Here are some suggestions:

Books:

- **"NFPA 70®: National Electrical Code (NEC) Handbook" by NFPA:** This comprehensive handbook provides in-depth explanations and interpretations of the National Electrical Code, helping electricians understand and apply the code in their work.
- **"Ugly's Electrical References" by George V. Hart:** A handy pocket-sized guide, Ugly's Electrical References offers quick access to essential electrical information, including formulas, conversions, and code-related details.

- **"Delmar's Standard Textbook of Electricity" by Stephen L. Herman:** This textbook covers fundamental electrical principles and serves as a comprehensive resource for understanding the basics of electrical theory and applications.
- **"Audel Electrical Course for Apprentices and Journeymen" by Paul Rosenberg:** Tailored for electricians at various stages of their careers, this book covers a wide range of topics, from basic electrical concepts to advanced skills needed by journeymen.

Online Resources:

- **National Electrical Contractors Association (NECA):** NECA provides resources, publications, and industry news, making it a valuable site for staying informed about industry trends, standards, and best practices.
- **Mike Holt Enterprises:** Mike Holt's website offers electrical training materials, books, videos, and online courses. It's a comprehensive resource for continuing education and exam preparation.
- **Electrical Code Academy, Inc.:** This online platform offers training courses and resources specifically designed to help electricians understand and apply the National Electrical Code.

Procurement Sites:

- **Grainger (grainger.com):** Grainger is a well-established supplier offering a wide range of electrical products, tools, and safety equipment. It's a go-to site for procuring quality materials.
- **Home Depot (homedepot.com) and Lowe's (lowes.com):** These retail giants provide a convenient way to procure electrical tools, materials, and equipment for both residential and commercial projects.
- **Graybar (graybar.com):** Graybar is a leading distributor of electrical, communications, and data networking products. It's a reliable source for procuring quality electrical supplies.

C. Comprehensive List of State Licensing Boards and Reciprocity Agreements

To Waive License Requirements in	You Need to Have a Current License in
Alabama	No Reciprocity Agreements
Alaska	Arkansas, Colorado, Iowa, Minnesota, Montana, Nebraska New Mexico, North Dakota, South Dakota, Oklahoma, Texas, Wyoming
Arizona	No Reciprocity Agreements
Arkansas	Alaska, New Mexico, Minnesota, Montana, New Hampshire,

	Oklahoma, Oregon, South Dakota Texas, Utah, Colorado, Oregon (For master license only)
California	No Reciprocity Agreements
Colorado	Alaska, Arkansas, Idaho, Iowa, Minnesota, Montana, Nebraska, New Hampshire, New Mexico, North Dakota, Oklahoma, South Dakota, Utah, Wyoming
Connecticut	No reciprocity agreements
Delaware	Approved on a case-by-case basis. Must be licensed in a jurisdiction that has substantially similar license requirements to Delaware.
Florida	No reciprocity agreements
Georgia	No Reciprocity Agreements
Hawaii	No reciprocity agreements
Idaho	Colorado, North Dakota, Maine, Oklahoma, Montana, Oregon, Nebraska, South Dakota, New Hampshire, Texas, New Mexico, Utah, Wyoming
Illinois	No reciprocity agreements
Indiana	No reciprocity agreements
Iowa	Alaska, Arkansas, Colorado, Minnesota, Montana, Nebraska, North Dakota, Oklahoma, South Dakota, Texas, Wisconsin
Kansas	No reciprocity agreements
Kentucky	Louisiana, Ohio, Virginia, West Virginia
Louisiana	No reciprocity agreements
Maine	Journeyman: New Hampshire, Vermont, North Dakota, Idaho, Oregon, Wyoming. Master: New Hampshire, Vermont
Maryland	Maryland, Washington, DC, West Virginia, Delaware. Requirements vary a lot by state, refer to website
Massachusetts	New Hampshire
Michigan	No reciprocity agreements
Minnesota	Journeyman: Alaska, Arkansas, Colorado, Iowa, Montana, Nebraska, North Dakota, South Dakota, Wyoming Master: Iowa, North Dakota, South Dakota, Nebraska
Mississippi	No reciprocity agreements
Missouri	No reciprocity agreements
Montana	Arkansas, Arizona, Colorado, Minnesota, North Dakota, Nebraska,

	New Hampshire, New Mexico, Oklahoma, South Dakota, Texas, Utah, Wyoming, Idaho
Nebraska	Alaska, Arkansas, Colorado, Iowa, Idaho, Minnesota, Montana, New Mexico, North Dakota, Oklahoma, South Dakota, Texas, Wyoming
Nevada	No reciprocity agreements
New Hampshire	Journeyman: Alaska, Arkansas, Colorado, Montana, North Dakota, South Dakota, Utah, Wisconsin, Wyoming, Iowa. Master: Maine, Massachusetts, Vermont
New Jersey	No reciprocity agreements
New Mexico	Alaska, Arkansas, Colorado, Idaho, Montana, Nebraska, Oklahoma, South Dakota, Texas, Utah, Wyoming
New York	No reciprocity agreements
North Carolina	No reciprocity agreements
North Dakota	Journeyman: Alaska, Colorado, Idaho, Iowa, Maine, Montana, Nebraska, New Hampshire, Utah, Wyoming Master: Minnesota, South Dakota
Ohio	No reciprocity agreements
Oklahoma	Alaska, Arkansas, Colorado, Idaho, Iowa, Montana, Nebraska, South Dakota, Texas, Wyoming
Oregon	Arkansas, Idaho, Maine, Montana, Utah, Washington, Wyoming
Pennsylvania	No reciprocity agreements
Rhode Island	No reciprocity agreements
South Carolina	No reciprocity agreements
South Dakota	Alaska, Arkansas, Colorado, Idaho, Iowa, Minnesota, Montana, Nebraska, New Hampshire, New Mexico, North Dakota, Oklahoma, South Dakota, Texas, Utah, Wyoming
Tennessee	No reciprocity agreements
Texas	Journeyman: Alaska, Arkansas, Idaho, Iowa, Montana, Nebraska, New Mexico, Oklahoma, South Dakota, Wyoming Master: North Carolina
Utah	Alabama, Alaska, Arkansas, California, Colorado, Connecticut, South Dakota, Delaware, New Hampshire, Hawaii, Idaho, Iowa,

	New Jersey, Kentucky, Maine, Massachusetts, New Mexico, Michigan, Minnesota, Montana, Nebraska, Oklahoma, Oregon, Rhode Island, Texas, Vermont, West Virginia, Virginia, Washington, Wisconsin, Wyoming
Vermont	Maine, New Hampshire
Virginia	Kentucky, Maryland, West Virginia, North Carolina
Washington	Oregon
West Virginia	All 50 States
Wisconsin	Iowa, New Hampshire
Wyoming	Journeyman: Alaska, Colorado, Idaho, Maine, Minnesota, Montana, Nebraska, New Hampshire, New Mexico, North Dakota, Oklahoma, Oregon, South Dakota, Texas, Iowa. Master: Idaho, South Dakota, Iowa

D. Glossary of Technical Terms and Key Concepts

Ampacity: Refers to the maximum continuous current, measured in amperes, that a conductor can safely carry under specified conditions without surpassing its temperature rating.

Bonding: Involves the permanent connection of metallic parts to establish an electrically conductive path, ensuring both electrical continuity and the ability to safely conduct any fault current that may be imposed.

Conduit: A tube or trough designed to safeguard electric wires or cables.

Dwelling Unit: A singular living space equipped with independent facilities for cooking, living, and sleeping.

Equipment Grounding Conductor: A conductor employed to link non-current-carrying metal components of equipment, raceways, and other enclosures to the system grounded conductor or grounding electrode conductor.

Fault Current: The current that flows when a fault occurs in an electrical system.

Ground Fault: An unintentional, electrically conductive connection between an ungrounded conductor and the grounding system or equipment grounding conductor.

Hazardous Location: An area where fire or explosion hazards may exist due to flammable gases or vapors, combustible dust, or ignitable fibers or flyings.

Interrupting Rating: The highest current at rated voltage that a device is identified to interrupt under standard test conditions.

Junction Box: An enclosure designed to house and protect electrical connections.

Kinetic Energy: The energy an object possesses due to its motion.

Listed: Equipment or materials included in a list published by an organization acceptable to the authority having jurisdiction and concerned with evaluation of products.

Motor Branch Circuit: A circuit that supplies energy to the motor control apparatus or motor load.

Neutral Conductor: A conductor that carries current under normal conditions and is intended to serve as a current-carrying conductor.

Overcurrent: Any current more than the rated current of equipment or the ampacity of a conductor.

Power Outlet: A point in the wiring system where current is taken to supply utilization equipment.

Qualified Person: Someone with expertise and proficiency in the design and functioning of electrical equipment and installations, coupled with training in safety protocols.

Raceway: An enclosed channel designed expressly for holding wires, cables, or busbars.

Service Entrance: The point where electric power is brought into a building or other structure and electrical equipment is installed to control and distribute power.

Transformer: A device consisting of two or more coils of wire used to transfer electrical energy by electromagnetic induction from one coil to another.

Ungrounded Conductor: A conductor that is not intentionally grounded.

Voltage Drop: The voltage reduction in an electrical circuit resulting from impedance.

Watt: The unit of electrical power, equal to one joule per second.

X-Rating: A design factor assigned to equipment for use in locations where the prevailing conditions are likely to cause an explosion.

Yoke: A device that combines two or more receptacles or switches in one mounting strap.

Zone: An area classified as potentially hazardous due to the presence of flammable gases or vapors.

Conclusion

Celebrating the Journey and Embracing Continuous Growth

Celebrating the journey as a Journeyman Electrician is not just about reaching milestones; it is about embracing the continuous growth inherent in this dynamic profession. Each successfully completed project, every troubleshooting triumph, and the mastery of new skills contribute to the rich tapestry of a journeyman's career. It is a celebration of resilience, adaptability, and the dedication to honing one's craft. Beyond the technical expertise gained, the journey instills a deep sense of responsibility for ensuring the safety and reliability of electrical systems. As a Journeyman Electrician, the commitment to lifelong learning becomes a cornerstone. Every new code edition, technological innovation, or advanced certification marks another opportunity for growth. It is about staying attuned to industry shifts, embracing emerging trends, and integrating sustainable practices. The journeyman's path is one of continuous improvement, not just professionally but personally, embodying the values of dedication, precision, and a profound respect for the power they wield. In celebrating the journey, the journeyman electrician recognizes that the true essence lies not just in where they've been, but in the exciting journey that lies ahead, ever evolving and filled with new challenges and triumphs.

Final Words of Encouragement and Inspiration

To the dedicated Journeyman Electrician, I offer these final words of encouragement and inspiration. Your journey is a testament to your perseverance, expertise, and unwavering commitment to the electrical craft. As you navigate the intricate circuits and systems, remember that every obstacle you've overcome has only fortified your capabilities. Your work is not merely about wires and circuits; it's about ensuring the safety and functionality of the spaces we inhabit. Embrace the evolving landscape of the electrical industry, for within it lies the opportunity for continual growth and innovation. Your skill, precision, and dedication illuminate a path for those who follow, and your impact extends far beyond the wires you connect. As you embark on each new project, carry with you the pride of a journey well-traveled and the excitement of challenges yet to be conquered. Your role as a Journeyman Electrician is not just a profession; it's a calling to illuminate the world with your expertise. So, stand tall amidst the conduits of opportunity, for your journey is an inspiration to aspiring electricians and a source of pride for the industry. May each connection you make be a testament to the excellence you bring to your craft. Continue to shine brightly, for you are the beacon of a profession that powers the world.

BONUS

To express our gratitude and to add even more value to your purchase, we've prepared exclusive BONUS content just for you. This treasure trove of materials is not static; it's a living, evolving resource designed to continuously enrich your experience. We encourage you to scan the QR code below to unlock your free bonus materials.

But that's not all. We believe in constantly improving and expanding the resources available to you. As such, we invite you to revisit this QR code periodically. Each visit could unveil new surprises, updated content, and even more in-depth insights to enhance your knowledge and enjoyment.

Consider this our way of saying thank you and to ensure that your investment keeps on giving. So, don't miss out on this opportunity to unlock additional value and keep abreast of all the fresh, exciting bonuses we have in store for you!

Scan the QR code now and dive into a world of added benefits and exclusive materials, tailor-made for our valued readers like you. Your adventure of discovery and learning continues beyond the pages of this book.

Download Here

www.ingramcontent.com/pod-product-compliance
Lightning Source LLC
Chambersburg PA
CBHW081103290526
45795CB00006B/1976